Synthesis Lectures on Renewable Energy Technologies

The series, Synthesis Lectures on Renewable Energy Technologies publishes concise books, focused on technologies that harness energy from naturally occurring sources, such as sunlight, wind, water, geothermal heat, and biofuels from organic materials. These renewable energy technologies play a crucial role in transitioning away from fossil fuels, helping to mitigate the effects of climate change, and promoting a sustainable energy supply.

Abdelkhalek Chellakhi · Said El Beid

Optimizing Solar Photovoltaic Systems

Advances in MPPT Techniques for Enhanced Energy Efficiency

Abdelkhalek Chellakhi ⓘ
Laboratory of Electronics, Signals, Systems,
and Informatics (LESSI)
Department of Physics
Faculty of Sciences Dhar El Mahraz
Sidi Mohamed Ben Abdellah University
Fez, Morocco

Said El Beid
CISIEV Team
Cadi Ayyad University
Marrakech, Morocco

ISSN 2690-5000   ISSN 2690-5019 (electronic)
Synthesis Lectures on Renewable Energy Technologies
ISBN 978-3-031-93282-3   ISBN 978-3-031-93283-0 (eBook)
https://doi.org/10.1007/978-3-031-93283-0

© The Editor(s) (if applicable) and The Author(s), under exclusive license to Springer Nature Switzerland AG 2025

This work is subject to copyright. All rights are solely and exclusively licensed by the Publisher, whether the whole or part of the material is concerned, specifically the rights of translation, reprinting, reuse of illustrations, recitation, broadcasting, reproduction on microfilms or in any other physical way, and transmission or information storage and retrieval, electronic adaptation, computer software, or by similar or dissimilar methodology now known or hereafter developed.
The use of general descriptive names, registered names, trademarks, service marks, etc. in this publication does not imply, even in the absence of a specific statement, that such names are exempt from the relevant protective laws and regulations and therefore free for general use.
The publisher, the authors and the editors are safe to assume that the advice and information in this book are believed to be true and accurate at the date of publication. Neither the publisher nor the authors or the editors give a warranty, expressed or implied, with respect to the material contained herein or for any errors or omissions that may have been made. The publisher remains neutral with regard to jurisdictional claims in published maps and institutional affiliations.

This Springer imprint is published by the registered company Springer Nature Switzerland AG
The registered company address is: Gewerbestrasse 11, 6330 Cham, Switzerland

If disposing of this product, please recycle the paper.

Acknowledgements ﷽

First and foremost, we express our deepest gratitude to Allah, whose guidance and support have enabled us to achieve the completion of this book.

We extend our heartfelt thanks to the Faculty of Sciences Dhar El Mahraz at Sidi Mohamed Ben Abdellah University and the Faculty of Sciences and Technology at Cadi Ayad University for their invaluable assistance in bringing this work to fruition.

Our profound appreciation also goes to our beloved family members, particularly our dear mothers, to whom we owe an immense debt of gratitude. Their unwavering love, prayers, unconditional support, and encouragement have been a constant source of strength and inspiration throughout this journey.

Competing Interests The authors have no competing interests to declare that are relevant to the content of this manuscript.

Contents

1 Introduction 1
  1.1 Solar Energy Trends in Morocco: Overview and Problem Description ... 1
  1.2 Book Objectives 5
  1.3 Book Organization 5
  1.4 Summary 6
  References 6

2 Literature Review 9
  2.1 Introduction 9
  2.2 Solar Power Energy and Its Applications 10
  2.3 Types of Installation of Solar PV System 11
    2.3.1 Solar PV Systems Connected to the Grid (On-Grid) 12
    2.3.2 Off-Grid or Standalone Solar PV Systems 13
  2.4 DC–DC Converters 14
    2.4.1 DC–DC Boost Converter 16
    2.4.2 DC–DC Buck Converter 18
    2.4.3 DC–DC Interleaved Boost Converter 19
    2.4.4 DC–DC Three-Level Boost Converter 19
  2.5 Maximum Power Point Tracking (MPPT) Algorithm 21
    2.5.1 Importance of the MPPT Algorithm in Solar PV Systems 21
    2.5.2 Essential Specifications of MPPT Algorithm Design 21
    2.5.3 Description and Classification of MPPT Algorithms in PV Systems 23
    2.5.4 Overview of the Classification of MPPT Algorithms 25
  2.6 Battery Technologies 34
    2.6.1 Overview of Battery Technologies 35
  2.7 Summary 36
  References 37

3	Modeling and Simulation of Standalone Solar Photovoltaic Systems	49
	3.1 Introduction	49
	3.2 Modeling of a Standalone Solar Photovoltaic System	52
	3.2.1 Modeling of a Solar Photovoltaic Module	52
	3.2.2 MPPT System Modeling	71
	3.2.3 MATLAB/Simulink Software-Based Modeling	77
	3.2.4 Proteus Software Based Modeling	78
	3.3 Simulation of a Standalone Solar PV System with Conventional MPPT Algorithms	79
	3.3.1 P&O MPPT Algorithm-Based Implementation	79
	3.3.2 INC MPPT Algorithm-Based Implementation	82
	3.4 Summary	88
	References	89
4	Principle and Simulation Investigation of The Newly Proposed MPPT Approaches	93
	4.1 Introduction	93
	4.2 First Proposed Algorithm: An Enhanced MPPT Approach for Temperature Variation	95
	4.2.1 Principle	96
	4.2.2 MATLAB/Simulink Implementation and Simulation Results	99
	4.3 Second Proposed Algorithm: A Novel MPPT Tactic with Fast Tracking Speed and Zero Oscillation	109
	4.3.1 Principle	110
	4.3.2 Implementation and Simulation Results in MATLAB/Simulink Environment	111
	4.3.3 Implementation and Simulation Results in Proteus Environment	120
	4.4 Third Proposed Algorithm: A Novel Adaptable Step Size Theta Approach (ASSTA)	122
	4.4.1 Principle	124
	4.4.2 MATLAB/Simulink Implementation and Simulation Results	127
	4.5 Fourth Proposed Algorithm: An Innovative MPPT Approach for Temperature Varying with Zero Fluctuation and Fast-Converging Speed	136
	4.5.1 Principle	136
	4.5.2 MATLAB/Simulink Implementation and Simulation Results	138
	4.6 Summary	145
	References	149

5	**Experimental Validation**		151
	5.1	Introduction	151
	5.2	Experimental Setup Prototype of the Used Standalone PV System	151
		5.2.1 PV Array	152
		5.2.2 DC-DC Boost Converter	153
		5.2.3 DSPACE DS1104 Controller Card	154
		5.2.4 Working Procedure of the Used Standalone PV System	158
	5.3	Experimental Validation Results of the Four Novel Proposed MPPT Approaches	160
		5.3.1 Experimental Results of the First Proposed MPPT Approach	160
		5.3.2 Experimental Results of the Second Proposed MPPT Approach	161
		5.3.3 Experimental Results of the Third Proposed MPPT Approach	163
		5.3.4 Experimental Results of the Fourth Proposed MPPT Approach	166
	5.4	Summary	167
	References		168
6	**Conclusion and Future Research**		169
	6.1	General Conclusion	169
	6.2	Future Research Directions	171
Appendices			175

Abbreviations and Symbols

Abbreviations

ADC	Analog to Digital Converters
AI	Artificial Intelligence
ANN	Artificial Neural Network
ASSTA	Adjustable Step Size Theta Approach
BSS	Battery Storage System
CC	Constant Current
CCM	Continuous Conduction Mode
CE	Change in Error
CI	Coupled Inductor
CSP	Concentrated Solar Power
DAC	Digital to Analog Converters
DC	Direct Current
DCM	Discontinuous Conduction Mode
DNI	Direct Normal Irradiation
DSP	Digital Signal processing
dSPACE	Digital Signal Processing and Control Engineering
E	Error
ESS	Energy Storage System
FF	Fill Factor
FLC	Fuzzy Logic Controller
FOCV	Fractional Open Circuit Voltage
FSCC	Fractional Short Circuit Current
GA	Genetic Algorithm
GMPP	Global Maximum Power Point
HC	Hill Climbing
HGHPC	High Gain High Power Converters

HIL	Hardware In the Loop
IBC	Interleaved Boost Converter
INC	Increment of Conductance
LGLPC	Low Gain Low Power Converters
Li-ion	Lithium-ion
LMPP	Local Maximum Power Point
MPP	Maximum Power Point
MPPT	Maximum Power Point Tracking
NB	Negative Big
NiCd	Nickel Cadmium
NREL	National Renewable Energy Laboratory
NS	Negative Small
P&O	Perturbation And Observation
PB	Positive Big
PFC	Power Factor Correction
PI	Proportional-Integral
PID	Proportional-Integral-Derivative
PS	Positive Small
PSB	Polysulfide Bromine
PSC	Partial Shading Conditions
PSO	Particle Swarm Optimization
PV	Photovoltaic
PWM	Pulse Width Modulation
RCC	Ripple Correlation Control
RCP	Rapid Control Prototyping
RES	Renewable Energy Sources
RTI	Real-Time Interface
s	Seconds
SC	Soft Competing
SMC	Sliding Mode Controller
SoC	State of Charge
STC	Standard Test Conditions
TLBC	Three Level Boost Converter
VSSINC	Variable Sep Size INC
W	Watts
ZE	Zero
ZEBRA	Zero Emission Battery Research Activities

Symbols

A	Dimensionless junction material factor
AC	Alternating current
C_{in}	Input Capacitance (F)
C_{out}	Output Capacitance (F)
$D(k)$	Duty cycle of the boost converter at instant (k)
dI_{pv}	Photovoltaic current variation (A)
dP_{pv}	Photovoltaic voltage variation (W)
dV_{pv}	Photovoltaic power variation (V)
E_g	Semiconductor band-gap energy of solar cell (eV)
f	Frequency (Hz)
G	Solar irradiation level (W/m^2)
I_{ph}	Photo generated current (A)
I_{MPP}	Current at Maximum Power Point (A)
I_o	Reverse saturation current (A)
$I_{pv}(k)$	Photovoltaic cell current at instant k (A)
I_s	Saturation current (A)
I_{sc}	Short circuit current (A)
I-V	Current versus voltage features of PV panel
I_{so}	Reverse saturation current at Tr (A)
K	Boltzmann's constant (8.617 × 10^{-5}) (eV/K)
K_i	Short-circuit current temperature coefficient
K_v	Open-circuit voltage temperature coefficient
L	Inductance (H)
N	Scaling factor
N_p	Number of parallel PV cell in a PV panel
N_s	Number of series PV cell in a PV panel
$P_{MPP}(k)$	Instantaneous extracted PV panel power at instant (k) (W)
$P_{MPPa}(k)$	Instantaneous available PV panel power at instant (k) (W)
P_{MPP}	Power at maximum power point (W)
P_{pv}	Photovoltaic cell output power (W)
P-V	Power versus voltage features of PV panel
q	Electronic charge (1.6 × 10^{-19}) (C)
R_L	Resistance of load charge (Ω)
R_p	Parallel resistance in model of solar cell (Ω)
R_s	Series resistance in model of solar cell (Ω)
T	Temperature of the p-n junction (K)
T_r	Reference temperature of solar cell (K)
V_{MPP}	Voltage at Maximum Power Point (V)
V_{oc}	Open circuit voltage (V)

$V_{pv}(k)$	PV cell output voltage at instant k (V)
V_t	Solar cell thermal voltage (V)
η_{MPPT}	Tracking efficiency of MPPT approach
$\eta_{MPPT(avg)}$	Average tracking efficiency of MPPT approach
ϑ	Angle between the tangent line of the photovoltaic array power-voltage curves and V-axis
$d\vartheta/dV_{pv}$	Derivation of angle θ concerning de photovoltaic array voltage

List of Figures

Fig. 1.1	RES installed capacity in the Kingdom of Moroccan	2
Fig. 1.2	Solar potential of the Moroccan Kingdom [13]	2
Fig. 1.3	Moroccan Kingdom Direct Normal Irradiation (DNI) map [13]	3
Fig. 1.4	Map of solar project locations in the Kingdom of Morocco [13]	4
Fig. 2.1	Annual solar PV installation capacity from 2000 to 2023 [6]	11
Fig. 2.2	PV system installation: **a** off-grid and **b** on-grid	12
Fig. 2.3	Transforming the energy scenario of renewable energy generation (TWh/year) [12]	13
Fig. 2.4	Applications of standalone PV panels in remote areas: **a** rooftops, **b** road lighting, **c** small devices, **d** traffic signalization, and **e** water pumping	14
Fig. 2.5	The general applications areas of standalone solar PV systems [4]	15
Fig. 2.6	Non-isolated DC–DC converters: **a** boost, **b** buck, **c** interleaved boost, and **d** three-level boost	16
Fig. 2.7	Operating states of the boost converter: **a** ON and **b** OFF	17
Fig. 2.8	Operating states of the buck converter: **a** ON and **b** OFF	18
Fig. 2.9	Operating modes of the interleaved boost converter (IB) circuit: **a** mode 1, **b** mode 2, **c** mode 3, and **d** mode 4	20
Fig. 2.10	Operating modes of the three-level boost converter (TLBC) circuit: **a** mode 1, **b** mode 2, **c** mode 3, and **d** mode 4	21
Fig. 2.11	P–V and I-V characteristics of a photovoltaic array	22
Fig. 2.12	Main specifications of MPPT design	22
Fig. 2.13	**a** Duty ratio-based and **b** current-/voltage-based MPPT approach implementation	24
Fig. 2.14	Flowcharts of the P&O MPPT algorithm: **a** Direct control method, **b** Indirect control method	25
Fig. 2.15	Sing of I_{PV}/V_{PV} and dP_{PV}/dV_{PV} at different position of **a** I–V and **b** P–V curves, respectively	26

Fig. 2.16	Flowchart of the voltage-based INC MPPT approach	27
Fig. 2.17	Flowchart of hill climbing (HC) MPPT algorithm	28
Fig. 2.18	Flowchart of the FSCC MPPT algorithm	28
Fig. 2.19	Flowchart of the FOCV MPPT algorithm	29
Fig. 2.20	Flowchart of the RCC MPPT algorithm	29
Fig. 2.21	A basic structure of the FLC MPPT approach	30
Fig. 2.22	A basic structure of the ANN MPPT approach	31
Fig. 2.23	Displacement of particles in the optimization process of the PSO algorithm	32
Fig. 2.24	Flowchart of the PSO MPPT algorithm	33
Fig. 3.1	Silicon (Si); **a** material and **b** crystalline structure of atoms	50
Fig. 3.2	A graphical representation of the operation of a PV cell	50
Fig. 3.3	Three types of PV cells based on Si material; **a** poly-crystalline, **b** mono-crystalline, and **c** thin-film amorphous	51
Fig. 3.4	Different types of standalone PV system installations: **a** unregulated, **b** regulated without batteries, and **c** regulated with batteries	52
Fig. 3.5	By-pass and anti-return diodes for PV module protection	53
Fig. 3.6	Configuration types and relationships between PV cell, module and array	54
Fig. 3.7	Equivalent solar cell electrical circuit; **a** ideal model, **b** one-diode model, **c** two-diode model, and **d** three-diode model	55
Fig. 3.8	Equivalent PV module electrical circuit with NS Series and NP parallel branches	57
Fig. 3.9	A graphical illustration of the main parameters of a PV module	58
Fig. 3.10	Block diagram of the PV module developed in the Simulink environment	59
Fig. 3.11	Contents of the PV module block	60
Fig. 3.12	Implementation of the photocurrent using Eq. (3.5). This corresponds to the subsystem labeled "subsystem of photocurrent" in Fig. 3.11	60
Fig. 3.13	Photocurrent implementation using Eq. (3.5). This represents the subsystem named "subsystem of photocurrent" in Fig. 3.11	61
Fig. 3.14	Diode reverse saturation current implementation using Eq. (3.6). This subsystem is referred to as the "subsystem of diode reverse saturation current" in Fig. 3.11	61
Fig. 3.15	Single-diode model implementation using Eq. (3.9). This subsystem is labeled "subsystem of single diode model" in Fig. 3.11	62
Fig. 3.16	Schematic diagram of the PV array block and its components	62
Fig. 3.17	Key parameters of the PV array block	63

Fig. 3.18	Pseudo-code of the MATLAB script for implementing the mathematical model of the PV module	63
Fig. 3.19	Block diagram of the PV module and its circuit components within the Proteus environment	64
Fig. 3.20	Circuit configuration used for generating the characteristic curves of the PV module in the Proteus environment	64
Fig. 3.21	Characteristics of the PV module under STC conditions: **a** I–V curve and **b** P–V curve	65
Fig. 3.22	P–V and I–V curves at varying solar intensity levels and a fixed temperature of 25 °C	66
Fig. 3.23	P–V and I–V curves at various temperature levels with constant solar insolation of 1 kW/m^2	67
Fig. 3.24	MPP variations with increasing temperature under different solar radiation levels	67
Fig. 3.25	**a** I–V and **b** P–V characteristics of the PV module under different partial shading conditions	68
Fig. 3.26	P–V and I–V curves at varying shunt resistance values	69
Fig. 3.27	P–V and I–V curves at different series resistance values	69
Fig. 3.28	P–V and I–V curves under various diode saturation current values	70
Fig. 3.29	P–V and I–V curves at different values of the ideality factor	70
Fig. 3.30	Coupling modes of the PV module and load in a standalone PV system: **a** Direct connection, **b** Via a DC-DC converter	71
Fig. 3.31	Operating point (OP) on the I–V and P–V curves of a PV module with a DC load	72
Fig. 3.32	Schematic circuit of the DC-DC boost converter	72
Fig. 3.33	Circuit modeling implementation of the boost converter using the Simulink platform	73
Fig. 3.34	Input and output voltage and current waveforms of the boost converter	73
Fig. 3.35	Schematic circuit of the buck converter	73
Fig. 3.36	Circuit modeling of the buck converter using the Simulink platform	74
Fig. 3.37	Input and output voltage and current curves of the buck converter	74
Fig. 3.38	Schematic circuit of the interleaved boost converter	75
Fig. 3.39	Circuit modeling of the IB converter using the Simulink platform	75
Fig. 3.40	Input and output waveforms of the interleaved boost converter	76
Fig. 3.41	Schematic circuit of the three-level boost converter	76
Fig. 3.42	Circuit modeling of the three-level boost converter using the Simulink platform	77
Fig. 3.43	Input and output voltage and current waveforms of the three-level boost converter	77

Fig. 3.44	MATLAB/Simulink model of the overall standalone solar PV system	78
Fig. 3.45	Proteus software model of the overall standalone solar PV system	79
Fig. 3.46	Diagram of the P&O MPPT algorithm	81
Fig. 3.47	Simulink implementation of the P&O MPPT algorithm using M-file code in an embedded MATLAB function	81
Fig. 3.48	Profile of solar irradiation variation	82
Fig. 3.49	PV output power curves obtained using the P&O MPPT algorithm in the Simulink environment	82
Fig. 3.50	Duty cycle curves obtained using the P&O MPPT algorithm in the Simulink environment	83
Fig. 3.51	PV voltage and current curves obtained using the P&O MPPT algorithm in the Simulink environment	83
Fig. 3.52	PV power versus voltage curves obtained using the P&O MPPT method in the Simulink environment	84
Fig. 3.53	Duty cycle versus voltage curves obtained using the P&O MPPT method in the Simulink environment	84
Fig. 3.54	Overall implementation of the standalone PV system using the P&O MPPT method in the Proteus environment	85
Fig. 3.55	PV output power waveforms obtained using the P&O MPPT method in the Proteus environment	85
Fig. 3.56	Diagram of the INC MPPT algorithm	86
Fig. 3.57	Implementation of the INC MPPT algorithm using M-file code in an embedded MATLAB function	86
Fig. 3.58	PV module output power tracking curves obtained using the INC MPPT method	87
Fig. 3.59	Duty cycle simulation curves obtained using the INC MPPT method	87
Fig. 3.60	PV voltage and current simulation curves obtained using the INC MPPT method	87
Fig. 3.61	Power versus voltage tracking simulation curves using the INC MPPT method	88
Fig. 3.62	Duty cycle versus voltage tracking simulation curves using the INC MPPT method	88
Fig. 3.63	PV output power waveforms obtained using the INC MPPT technique in the Proteus environment	89
Fig. 4.1	A schematic overview of the block diagram for a standalone PV system equipped with an MPPT controller	94
Fig. 4.2	Schematic representation of the key challenges faced by conventional MPPT methods	94

Fig. 4.3	Illustration of the changes in the MPP on the P-V curves in response to temperature variations at different levels of solar radiation	95
Fig. 4.4	The movement of the OP during a rapid increase in temperature	97
Fig. 4.5	The movement of the OP during a rapid decrease in temperature	97
Fig. 4.6	A detailed view of the OP movement procedure during a rapid increase in temperature	98
Fig. 4.7	A detailed view of the OP movement procedure during a rapid decrease in temperature	98
Fig. 4.8	Flowchart of the proposed enhanced MPPT approach	99
Fig. 4.9	Circuit of the proposed standalone PV system with the suggested MPPT approach	100
Fig. 4.10	Temperature variation profile	101
Fig. 4.11	Simulation results of the improved MPPT approach (P&O-IMP) compared to the traditional P&O MPPT algorithm under temperature variation for fixed irradiance levels: **a** 1 kW/m^2 and **b** 0.6 kW/m^2	102
Fig. 4.12	Steady-state error analysis of power and voltage under STC	103
Fig. 4.13	Comparison of response time under STC	103
Fig. 4.14	Results of tracking efficiency for the improved MPPT approach (P&O-IMP) compared to the traditional P&O algorithm under temperature variation for fixed irradiance levels: **a** 1 kW/m^2 and **b** 0.6 kW/m^2	104
Fig. 4.15	Simulation results of the improved MPPT approach (INC-IMP) compared to those of the INC MPPT technique with regard to the temperature variability under fixed irradiation: **a** 1 kW/m^2 and **b** 0.6 kW/m^2	105
Fig. 4.16	Results of the tracking efficiency of the improved MPPT approach (INC-IMP) compared to the INC MPPT technique regarding the variability of temperature under fixed irradiation: **a** 1 kW/m^2 and **b** 0.6 kW/m^2	106
Fig. 4.17	Simulation results of the improved MPPT approach (Mod-MPP-Locus-IMP) compared to the Mod-MPP-Locus MPPT technique regarding temperature varying under fixed irradiation: **a** 1 kW/m^2 and **b** 0.6 kW/m^2	107
Fig. 4.18	Results of tracking efficiency of the improved MPPT approach (Mod-MPP-Locus-IMP) compared to the Mod-MPP-Locus MPPT technique regarding temperature varying under fixed irradiation: **a** 1 kW/m^2 and **b** 0.6 kW/m^2	108
Fig. 4.19	MPP voltage area illustration on P-V curves	110
Fig. 4.20	Flowchart of the second proposed MPPT approach [15]	111

Fig. 4.21	Simulink implementation of the complete standalone PV system	113
Fig. 4.22	Solar irradiance profiles: **a** sudden insolation changes and **b** sinusoidal insolation changes	114
Fig. 4.23	PV power tracked by different MPPT techniques under **a** sudden and **b** sinusoidal test conditions	115
Fig. 4.24	PV current, PV voltage, and duty cycle of different MPPT techniques under **a** sudden and **b** sinusoidal test conditions	116
Fig. 4.25	Tracking performance of the proposed approach: **a** P-V curve and **b** D-V curve under sudden insolation changes	117
Fig. 4.26	Tracking performance of the P&O method: **a** P-V curve and **b** D-V curve under sudden insolation changes	117
Fig. 4.27	Tracking performance of the INC method: **a** P-V curve and **b** D-V curve under sudden insolation changes	117
Fig. 4.28	Tracking performance of the proposed approach: **a** P-V curve and **b** D-V curve under sinusoidal insolation changes	118
Fig. 4.29	Tracking performance of the P&O method: **a** P-V curve and **b** D-V curve under sinusoidal insolation changes	118
Fig. 4.30	Tracking performance of the INC method: **a** P-V curve and **b** D-V curve under sinusoidal insolation changes	118
Fig. 4.31	Load profile test under varying load conditions	119
Fig. 4.32	PV power waveforms tracked by MPPT techniques under load variation test conditions	119
Fig. 4.33	Duty cycle curves of MPPT techniques under load variation test conditions	119
Fig. 4.34	Implementation of the entire standalone PV system using the Proteus environment	121
Fig. 4.35	Insolation profile; **a** case 1 and **b** case 2	122
Fig. 4.36	PV power waveforms of MPPT techniques under variation of insolation conditions; **a** case 1 and **b** case 2	123
Fig. 4.37	P-V characteristics under change in weather conditions; **a** temperature and **b** solar irradiance and **c** PSC	124
Fig. 4.38	The geometry of the Theta (θ) angle and its derivation $d\theta/dVpv$ diagram in the P-V characteristics under normal climatic conditions	125
Fig. 4.39	Characteristics of P-V, I-V and $(dPpv + dVpv * dIpv)$-V under uniform climatic conditions	126
Fig. 4.40	Graphical illustration of MPPT operation in case of using a fixed and adjustable perturbation step size	127
Fig. 4.41	Flow chart of the novel ASSTA MPPT approach	128
Fig. 4.42	The entire PV system implementation using the Simulink environment of MATLAB software	129

Fig. 4.43	PV array configurations	130
Fig. 4.44	Comparison of the PV power and duty cycle waveforms of the novel ASSTA with that of the INC, P&O, VSSINC and PSO MPPT strategies under; **a** STC and **b** sudden insolation variation conditions	131
Fig. 4.45	Comparison of the novel ASSTA with INC, P&O, VSSINC, and PSO strategies under fast temperature change scenario; **a** PV power, and **b** duty cycle waveforms	133
Fig. 4.46	Comparison of PV power and duty ratio waveforms of ASSTA with INC, P&O, VSSINC, and PSO under PSC; **a** Case 2, **b** Case 3	135
Fig. 4.47	Illustration of the MPP current zone (CZ) in the I-P and I-V characteristics of the PV module	137
Fig. 4.48	Flow chart of the innovative MPPT scheme	138
Fig. 4.49	The first scenario of temperature varying	139
Fig. 4.50	The second scenario of temperature varying	139
Fig. 4.51	Simulation results comparing the proposed MPPT with INC and P&O under high irradiance (1 kW/m^2) for perturbation step sizes of 0.5 and 0.1%	140
Fig. 4.52	Tracking efficiency of the MPPT methods under high irradiance and the first temperature scenario	141
Fig. 4.53	Simulation results comparing the proposed MPPT with INC and P&O under low irradiance (0.2 kW/m^2) for perturbation step sizes of 0.5 and 0.1%	142
Fig. 4.54	Tracking efficiency of the MPPT methods under low irradiance and the first temperature scenario	143
Fig. 4.55	Simulation results for the proposed MPPT method compared to INC and P&O under high irradiance (1 kW/m^2) with 0.1 and 0.5% perturbation step sizes	144
Fig. 4.56	Tracking efficiency of MPPT methods under high irradiance in the second temperature scenario	145
Fig. 4.57	Simulation results for the proposed MPPT method compared to INC and P&O under low irradiance (0.2 kW/m^2) with 0.1 and 0.5% perturbation step sizes	146
Fig. 4.58	Tracking efficiency of MPPT methods under low irradiance in the second temperature scenario	146
Fig. 5.1	Experimental prototype setup of the proposed standalone PV system developed in the CISIEV Team Laboratory	152
Fig. 5.2	Experimental results of the P-V and I-V characteristics of the PV array used	154

Fig. 5.3	Real solar irradiance and ambient temperature conditions for two days in the city of Marrakech; **a** 11 and **b** 12 August, 2022	155
Fig. 5.4	Conventional method for plotting the I-V characteristics of the PV array	156
Fig. 5.5	A screenshot of the dSPACE ControlDesk environment constructed for plotting P-V and I-V characteristics	156
Fig. 5.6	Illustration of the principal components of the DC-DC boost converter used in the practical validation	157
Fig. 5.7	Illustration of the hardware components of the DS1104 R&D controller board	157
Fig. 5.8	Illustration of the interface of the dSPACE ControlDesk program	158
Fig. 5.9	Schematic diagram explaining the connections between the components of the experimental prototype of the proposed photovoltaic system	159
Fig. 5.10	Screenshot of the implementation diagram of the MPPT algorithm in the Simulink environment using the dSPACE interface	159
Fig. 5.11	Experimental results of the IMP-P&O MPPT compared to the traditional P&O method	161
Fig. 5.12	Experimental results of the IMP-INC MPPT compared to the traditional INC method	162
Fig. 5.13	Architecture of the used PV system to emulate a fast insolation change situation	162
Fig. 5.14	Experimental results of the novel MPPT approach compared to INC and P&O techniques under normal conditions	163
Fig. 5.15	Experimental results of the novel MPPT approach compared to INC and P&O techniques under rapid insolation variations, including **a** PV power and duty cycle waveforms and **b** PV current and voltage waveforms	164
Fig. 5.16	Experimental results of the ASSTA MPPT approach compared to INC and P&O techniques under normal conditions	165
Fig. 5.17	Experimental results of the ASSTA approach compared to INC and P&O techniques under rapid insolation variation, including **a** PV power and duty cycle waveforms and **b** PV current and voltage waveforms	166
Fig. 5.18	Experimental results of the fourth new MPPT approach compared to INC and P&O algorithms under stable conditions	167

List of Tables

Table 2.1	Description of IB converter operating modes [38]	19
Table 3.1	Parameters of equivalent electrical circuit models of a PV cell and their meaning	57
Table 3.2	Parameters of the 1Soltech 1STH-215-P PV module under STC	65
Table 3.3	Ideality factor for different PV technologies	70
Table 3.4	Parameters used in the implementation of the standalone solar PV system in MATLAB/Simulink and Proteus	80
Table 4.1	Parameters of the complete standlone PV system (PV module (1Soltech 1STH-215-P) in the STC and DC-DC boost converter)	101
Table 4.2	Summarize of performances of the improved MPPT approaches compared to the conventional ones	109
Table 4.3	Average tracking efficiency comparison of MPPT methods under MATLAB/Simulink environment	120
Table 4.4	Performance summary of different MPPT strategies based on simulation results	120
Table 4.5	Parameters of the complete photovoltaic system (MSX-60 PV module) in the STC	130
Table 4.6	Summary of the performance comparison of the ASSTA, P&O, INC, VSSINC, and PSO MPPT techniques under the STC, Case 2, and Case 3 of PSC scenarios	137
Table 4.7	Summary of tracking efficiencies for INC, P&O, and the proposed MPPT strategy under various simulation conditions	147
Table 5.1	Parameters used in the practical test bench of the proposed off-grid PV system	153

1 Introduction

1.1 Solar Energy Trends in Morocco: Overview and Problem Description

Amidst the pressing environmental challenges posed by widespread reliance on fossil fuels for energy, the global integration of renewable energy sources represents a pivotal strategy in mitigating these issues, particularly the reduction of greenhouse gas emissions. Recognizing this imperative, many developing countries have accelerated their adoption of renewable energy technologies within their electricity generation frameworks, anticipating substantial increases in renewable energy capacity in the coming years [1, 2].

Among the nations striving to advance their energy sectors, the Kingdom of Morocco has emerged as a strategic partner for the European Union in electricity production and a promising hub for energy exports. This positioning is largely attributed to Morocco's advantageous geographic location and climate, both of which support the extensive deployment of renewable energy sources (RES). With abundant solar and wind resources, Morocco is well-poised to become one of the world's leading producers of green hydrogen, with the potential to meet over 4% of global demand by 2030 [3, 4]. This vast potential represents a significant opportunity for Morocco to decarbonize its economy, generate wealth, and create employment opportunities [5].

At the COP 21 conference in Paris, Morocco committed to ambitious renewable energy targets, aiming for renewables to account for 52% of its total installed electricity generation capacity by 2030 (see Fig. 1.1) [6]. By leveraging its renewable energy capabilities, Morocco envisions itself as not only reducing its carbon footprint but also enhancing its socio-economic development, further solidifying its role as a key player in the global energy landscape [7].

Morocco's geographic features, such as its extensive insolation and high Direct Normal Irradiation (DNI) levels, position it prominently in global solar and wind energy rankings (Figs. 1.2 and 1.3). The country benefits from an average daily incident irradiance of

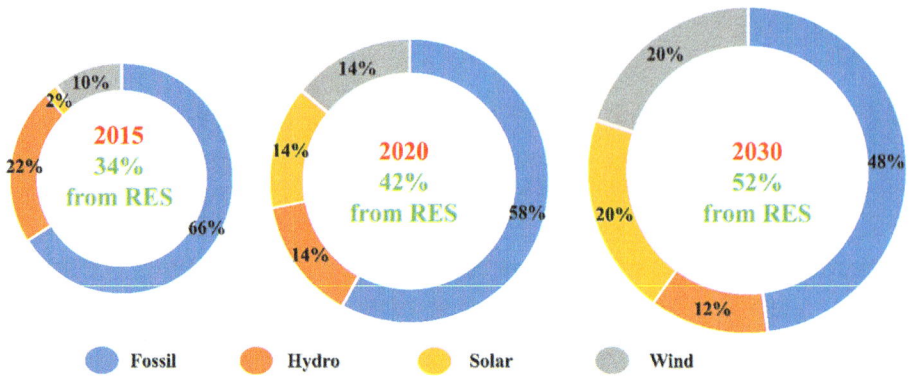

Fig. 1.1 RES installed capacity in the Kingdom of Moroccan

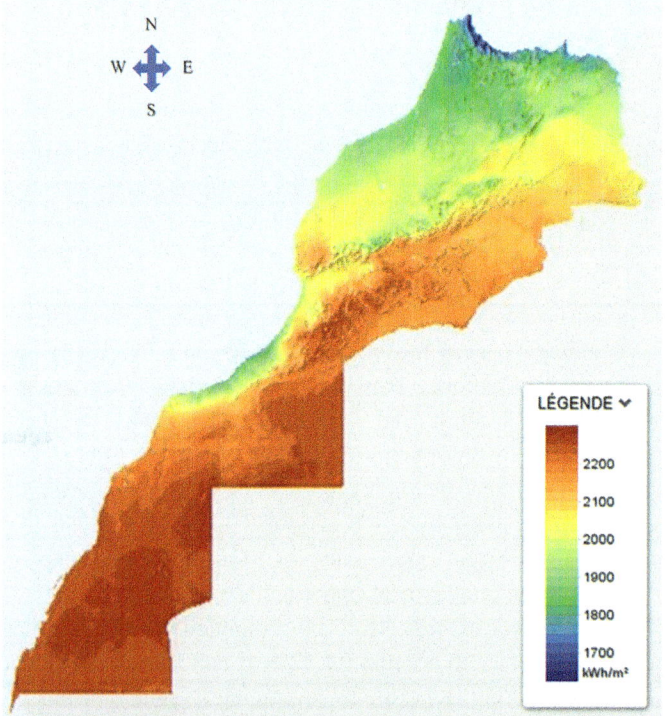

Fig. 1.2 Solar potential of the Moroccan Kingdom [13]

1.1 Solar Energy Trends in Morocco: Overview and Problem …

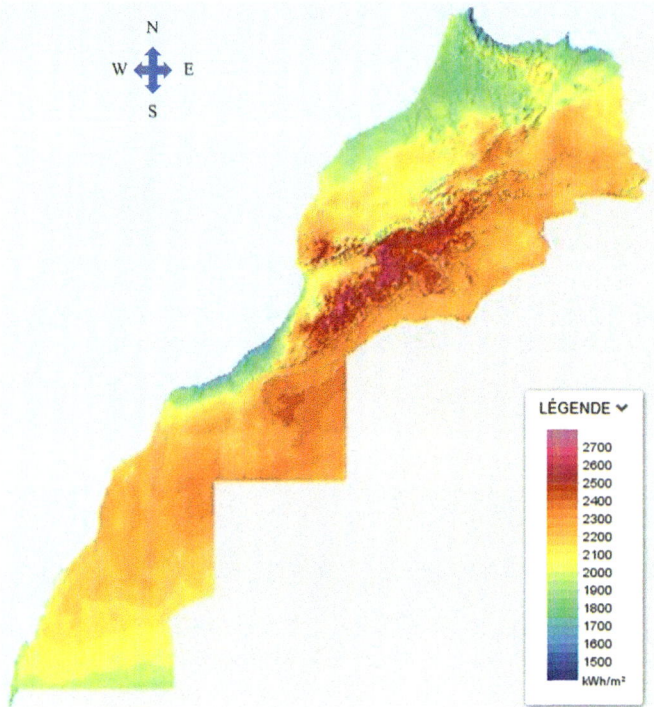

Fig. 1.3 Moroccan Kingdom Direct Normal Irradiation (DNI) map [13]

5.3 kWh/m_2 and experiences varying annual sunshine hours ranging from 2700 in the North to approximately 3500 in the South [6, 8]. These climatic attributes underscore Morocco's potential as a leading player in solar energy, supported by an ambitious solar energy plan. This initiative, one of the largest globally with an estimated investment of 8.4 billion dollars (7.8 billion euros), emphasizes Morocco's commitment to sustainable energy development [9].

Significant strides have already been made with multiple large-scale solar projects operational across various Moroccan cities. Notable projects include Ouarzazate, Ain Beni Mather, Foum Al Oued, and Boujdour, boasting substantial capacities of 580 MW, 472 MW, 500 MW, and 100 MW, respectively (Fig. 1.4) [6, 9]. These initiatives underscore Morocco's proactive approach towards achieving energy independence and meeting international climate commitments.

Photovoltaic (PV) technology stands at the forefront of Morocco's renewable energy efforts, characterized by its low maintenance requirements, portability, reliability, and decreasing costs. Despite these advantages, commercial PV panels typically exhibit conversion efficiencies below 25% [10]. This inefficiency is exacerbated by the nonlinear characteristics of PV panels' current-voltage (I–V) and power-voltage (P–V) curves,

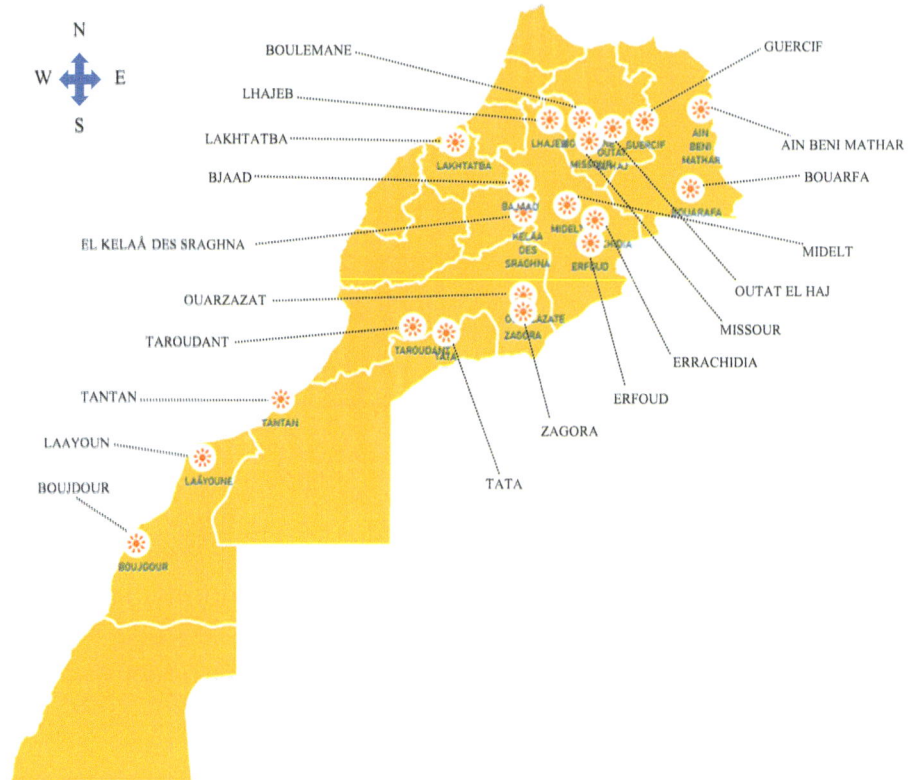

Fig. 1.4 Map of solar project locations in the Kingdom of Morocco [13]

which fluctuate significantly with environmental conditions such as solar irradiance, temperature variations, and load demands [11]. Maximizing PV panel efficiency hinges on precise tracking of the Maximum Power Point (MPP), a complex task addressed by advanced Maximum Power Point Tracking (MPPT) mechanisms [12].

In conclusion, Morocco's comprehensive solar energy strategy exemplifies its leadership in renewable energy adoption and underscores the critical role of innovative technologies in enhancing energy efficiency and sustainability. This book explores the pivotal role of MPPT techniques in optimizing PV system performance, providing insights and solutions crucial for advancing Morocco's renewable energy objectives and global sustainable development efforts.

1.2 Book Objectives

As we contemplate future energy solutions, it is crucial to consider how to power remote sites cost-effectively while minimizing greenhouse gas emissions. In this context, this book provides modeling and optimization strategies for standalone solar PV systems to supply isolated locations with safe, clean, renewable, and sustainable electricity. The primary focus of this work was to enhance the overall performance and efficiency of standalone solar PV systems by employing effective and robust MPPT approaches. These approaches ensure optimal transfer of the maximum available power to the load, especially during significant variations in weather conditions, such as abrupt changes in solar irradiation and temperature.

1.3 Book Organization

This book explores the advancements in solar PV systems, focusing on modeling, simulation, and optimization techniques to enhance energy efficiency. It covers innovative approaches to MPPT and their practical applications, providing insights into Morocco's renewable energy journey. Below is a summary of the chapters.

This chapter introduces Morocco's renewable energy initiatives, highlighting solar energy as a key pillar in achieving 52% renewable energy capacity by 2030. It emphasizes the role of PV technology and the challenges posed by environmental factors, necessitating advanced MPPT mechanisms.

Chapter 2 reviews the fundamentals of solar PV systems, types of direct current to direct current (DC-DC) converters, and classifications of MPPT algorithms—conventional, AI-based, and hybrid. It also discusses battery technologies relevant to PV systems.

Chapter 3 focuses on the modeling and simulation of standalone solar PV systems using MATLAB/Simulink and Proteus. It examines PV module characteristics, DC-DC converters, and conventional MPPT techniques like INC and P&O.

Chapter 4 presents four novel MPPT strategies designed to overcome the limitations of conventional methods. Their effectiveness is validated through simulations, showcasing superior tracking accuracy.

Chapter 5 provides the experimental setup and practical validation of the proposed MPPT strategies. The results confirm their applicability and performance under real-world conditions.

Chapter 6 concludes the book with key findings and directions for future research in PV system optimization.

1.4 Summary

This chapter introduces the growing role of solar energy in Morocco's renewable energy strategy. As a key player in the global energy market, Morocco has prioritized solar and wind power, leveraging its favorable geographic and climatic conditions. The country's ambitious renewable energy targets, including a commitment to generating 52% of its electricity from renewable sources by 2030, underscore its dedication to reducing carbon emissions while fostering socio-economic development. This chapter highlights Morocco's abundant solar resources, particularly its high levels of Direct Normal Irradiation (DNI), and examines its large-scale solar projects, which position the country as a significant contributor to global solar energy production.

Despite the advantages of photovoltaic (PV) technology—such as low maintenance requirements and declining costs—PV systems often suffer from inefficiencies due to fluctuating environmental conditions. The chapter emphasizes the critical role of advanced Maximum Power Point Tracking (MPPT) techniques in optimizing PV panel efficiency, thereby maximizing energy extraction. This introduction lays the groundwork for exploring innovative technologies that enhance PV system performance and support Morocco's renewable energy objectives.

References

1. G. Papaefthymiou, K. Dragoon, Towards 100% renewable energy systems: uncapping power system flexibility. Energy Policy **92**, 69–82 (2016). https://doi.org/10.1016/J.ENPOL.2016.01.025
2. A.A. Elbaset, *Performance Analysis of Photovoltaic Systems with Energy Storage Systems*
3. B.E. Lebrouhi, B. Lamrani, Y. Zeraouli, T. Kousksou, Key challenges to ensure Morocco's sustainable transition to a green hydrogen economy. Int. J. Hydrogen Energy **49** (2024). https://doi.org/10.1016/j.ijhydene.2023.09.178
4. S.R. Ersoy et al., Industrial and infrastructural conditions for production and export of green hydrogen and synthetic fuels in the MENA region: insights from Jordan, Morocco, and Oman. Sustain. Sci. **19**(1) (2024). https://doi.org/10.1007/s11625-023-01382-5
5. O. Bayssi et al., Green hydrogen landscape in North African countries: strengths, challenges, and future prospects. Int. J. Hydrogen Energy **84**, 822–839 (2024). https://doi.org/10.1016/j.ijhydene.2024.08.277
6. M. Boulakhbar et al., Towards a large-scale integration of renewable energies in Morocco. J. Energy Storage **32**, 101806 (2020). https://doi.org/10.1016/J.EST.2020.101806
7. R. Benbba et al., Solar energy resource and power generation in Morocco: current situation, potential, and future perspective. Resources **13**(10) (2024). https://doi.org/10.3390/resources13100140
8. T. Kousksou, A. Allouhi, M. Belattar, A. Jamil, T. El Rhafiki, Y. Zeraouli, Morocco's strategy for energy security and low-carbon growth. Energy **84**, 98–105 (2015). https://doi.org/10.1016/J.ENERGY.2015.02.048

References

9. A. Šimelytė, Promotion of renewable energy in Morocco, in *Energy Transformation Towards Sustainability* (2020), pp. 249–287. https://doi.org/10.1016/B978-0-12-817688-7.00013-6
10. M.B. Hayat, D. Ali, K.C. Monyake, L. Alagha, N. Ahmed, Solar energy—a look into power generation, challenges, and a solar-powered future. Int. J. Energy Res. **43**(3) (2019). https://doi.org/10.1002/er.4252
11. A.M. Eltamaly, A.Y. Almoataz, *Modern Maximum Power Point Tracking Techniques for Photovoltaic Energy Systems* (Springer, 2020)
12. M.G. Batarseh, M.E. Za'ter, Hybrid maximum power point tracking techniques: a comparative survey, suggested classification and uninvestigated combinations. Solar Energy **169** (2018). https://doi.org/10.1016/j.solener.2018.04.045
13. Mazen, Atlas de la Ressource Solaire au Maroc. https://solaratlas.masen.ma/map?c=28.902397:-9.074707:5&s=37.020098:-4.108887&m=masen:dni. Accessed 26 Mar 2025

Literature Review

2.1 Introduction

In a typical PV system, the overall energy conversion efficiency is influenced by the efficiency of both the PV panel and the components comprising the power conversion system. The efficiency of the PV panel itself is determined during manufacturing and remains fixed thereafter. However, the efficiency of the power conversion components, including the MPPT algorithm, DC–DC converter, and battery system, can be optimized and improved to enhance the overall system efficiency.

In standalone (off-grid) solar PV systems, the efficiency of the power conversion system is particularly critical due to the system's autonomy and reliance on stored energy. Therefore, selecting and integrating these components effectively is crucial for maximizing energy utilization and system reliability.

This chapter serves as a comprehensive literature review of the power conversion system in standalone solar photovoltaic systems. It begins by providing a thorough introduction to solar energy, detailing its applications and the various types of solar PV systems employed. The review then explores in-depth the essential components of the power conversion system:

- PV Panel: The foundational component that converts sunlight into electrical energy. Its efficiency is determined by the quality of materials and manufacturing processes.
- DC–DC Converters: Various types such as boost converters, buck converters, and their variants are discussed, focusing on their operational principles, advantages, and limitations in the context of solar PV systems.
- Battery Technologies: Highlighting different battery chemistries, such as lead-acid, lithium-ion, and their suitability for solar energy storage based on factors like energy density, cycle life, and cost-effectiveness.

- MPPT Algorithms: A critical aspect for maximizing the PV system's energy yield by continuously adjusting the operating point of the PV array to the MPP. Common algorithms like Perturbation and Observation (P&O), Increment of Conductance (INC), and newer adaptive techniques are reviewed, emphasizing their effectiveness and challenges in real-world applications.

This comprehensive review aims to provide foundational knowledge for researchers, engineers, and designers involved in solar PV system development. By understanding the strengths and limitations of current technologies, stakeholders can identify opportunities for innovation and improvement. Importantly, this chapter identifies gaps in existing MPPT algorithms, paving the way for subsequent research detailed in this book. Future chapters will focus on innovative MPPT strategies aimed at addressing these gaps to enhance the overall performance and efficiency of standalone solar PV systems.

2.2 Solar Power Energy and Its Applications

Solar energy is an abundant, inexhaustible, clean, and globally accessible resource, available in both direct (sun irradiation) and indirect forms (wind, biomass, ocean surface waves, etc.). It stands as a crucial pillar in combating global warming, reducing carbon dioxide (CO_2) emissions, and striving to maintain global temperatures below a 2 °C increase [1]. The International Energy Agency (IEA) report on global CO2 emissions in 2023 indicates that while clean energy growth helped limit the rise in global emissions to 1.1% that year, weather conditions and the post-COVID-19 economic recovery affected emission levels [2]. The report notes that advanced economies saw emissions decline to levels not observed in over 50 years, demonstrating the impact of accelerated clean energy deployment.

The energy from the Sun can manifest as either light or heat. There are various methods to harness this immense energy, with the most prevalent being its conversion into different energy forms such as heat, fuels, and electricity [3].

In the conversion of sunlight into electrical energy, two primary solar system technologies are most common: direct and indirect. Indirect solar system technology, known as solar thermal systems, utilizes concentrated solar power (CSP) to heat water, air, or other fluids. Conversely, direct solar system technology relies on photovoltaic systems, which generate electricity by converting sunlight into electrical energy via the photovoltaic effect. Due to its cost-effectiveness, ease of deployment, minimal maintenance requirements, and environmental friendliness, photovoltaic technology has emerged as the predominant method for exploiting solar energy. This technology finds extensive applications across a wide power spectrum, ranging from milliwatts to megawatts, including sectors such as transportation, agriculture, water pumping, telecommunications, and lighting [3, 4].

2.3 Types of Installation of Solar PV System

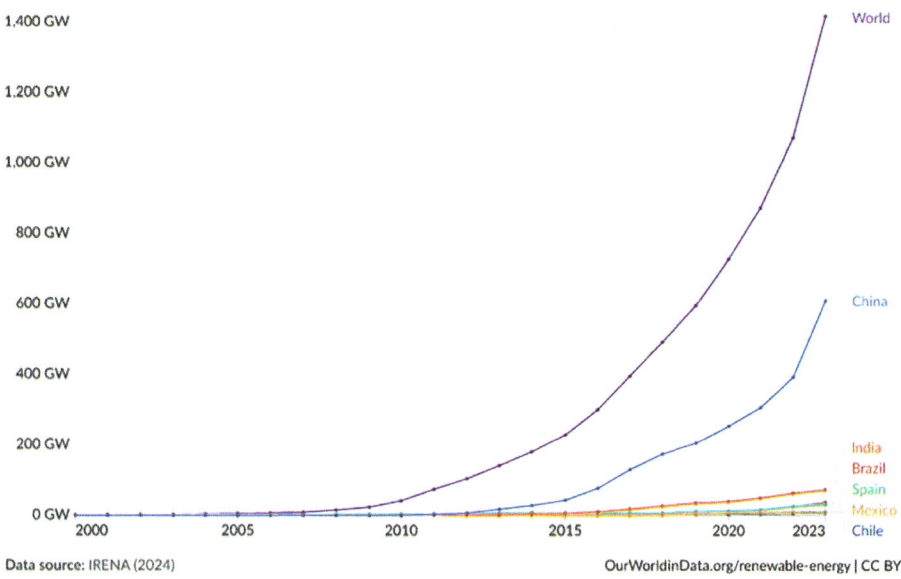

Fig. 2.1 Annual solar PV installation capacity from 2000 to 2023 [6]

In 2023, solar power led the global power generation capacity expansion, solidifying its position as the fastest-growing renewable energy source for the 19th consecutive year. Out of 576 GW of new renewable capacity added last year, solar PV contributed 78%, with 447 GW connected to the grid. This new solar capacity set a record, significantly surpassing analysts' projections with an impressive 87% growth rate (see Fig. 2.1). For comparison, 2022 saw the addition of 239 GW, reflecting a 46% year-on-year growth rate [5, 6].

2.3 Types of Installation of Solar PV System

Due to significant advancements in solar PV technology and continually decreasing prices, solar PV energy has become increasingly attractive over the last decade. There are three primary types of solar PV system installations: standalone (off-grid), grid-connected (on-grid), and hybrid systems [7]. Standalone (isolated) and on-grid systems are well-known types of solar PV installations. Basic schematics of standalone and grid-connected PV systems are shown in Fig. 2.2.

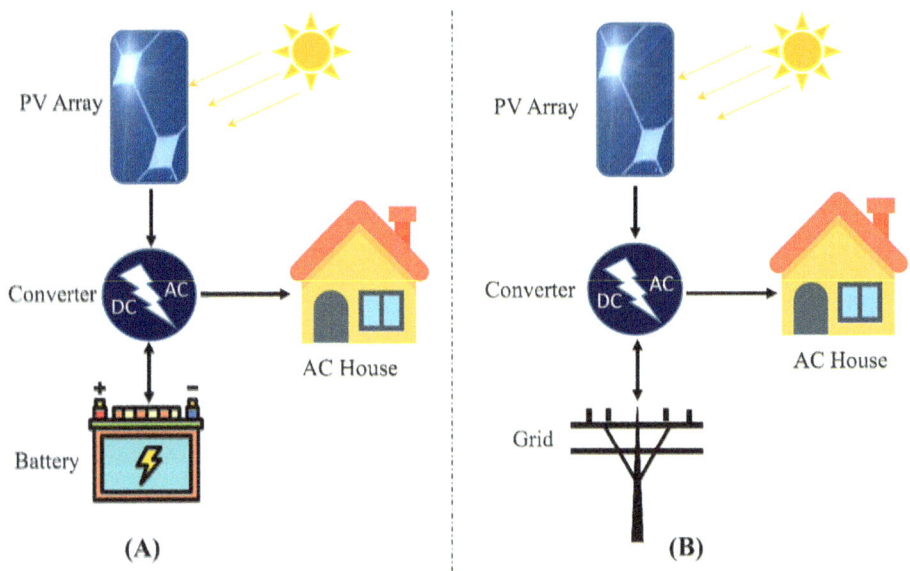

Fig. 2.2 PV system installation: **a** off-grid and **b** on-grid

2.3.1 Solar PV Systems Connected to the Grid (On-Grid)

On-grid solar PV systems are commonly found in both urban and rural areas of developed countries, with a higher concentration in urban areas. In rural regions of developing countries, on-grid installations are less common due to the lack of public electricity networks and the necessary infrastructure. However, urban areas in developing countries have seen a noticeable increase in on-grid PV systems, albeit with modest progress compared to developed nations [8]. On-grid solar PV systems have developed significantly but still face many challenges, as detailed in references [1, 9].

There are two main types of on-grid solar PV installations: with and without battery storage systems (BSS) [1].

- On-Grid PV Systems Without BSS: These installations are simple, inexpensive, require less maintenance, and have high efficiency. They can use the grid as a virtual battery, supplying excess generated electricity to the grid during peak periods. When solar energy is not available, such as at night or on cloudy days, the needed power can be drawn from the grid [1, 4].
- On-Grid PV Systems With BSS: These installations are more complex, expensive, and require more maintenance. However, they offer advantages such as selling excess energy during both peak and off-peak times and meeting load demands more effectively [1].

2.3 Types of Installation of Solar PV System

2.3.2 Off-Grid or Standalone Solar PV Systems

According to the World Bank Group report [10], global electricity access faced a significant setback in 2022, as the number of people living without power grew for the first time in over a decade. The latest figures show that 685 million individuals globally now lack access to electricity, a 10 million person increase from 2021. This reversal in progress is particularly pronounced in sub-Saharan Africa, which accounts for over 80% of the worldwide deficit, or 570 million people without access. This worrying trend represents a step backwards from the gradual improvements made in the region since 2010.

Due to recent technological advancements, renewable energy resources are expected to dominate global electricity generation by 2050, with solar PV energy playing a significant role Fig. 2.3 [11, 12]. Solar PV power is particularly important in remote areas that cannot connect to the grid (see Fig. 2.4). For these areas, off-grid PV systems are the best choice due to their reliability, portability, and ease of implementation. Off-grid systems have extensive applications (show Fig. 2.5), making them a viable electrification solution and a key component in achieving SDG7 [8].

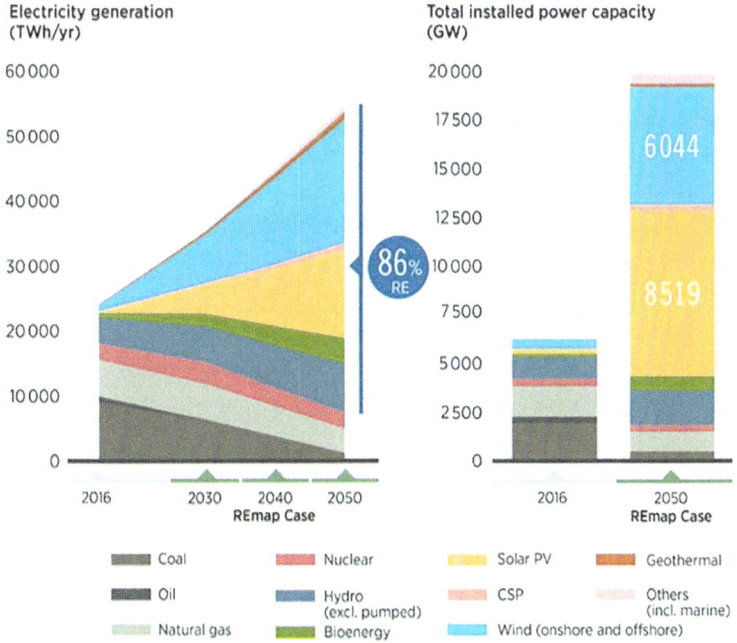

Fig. 2.3 Transforming the energy scenario of renewable energy generation (TWh/year) [12]

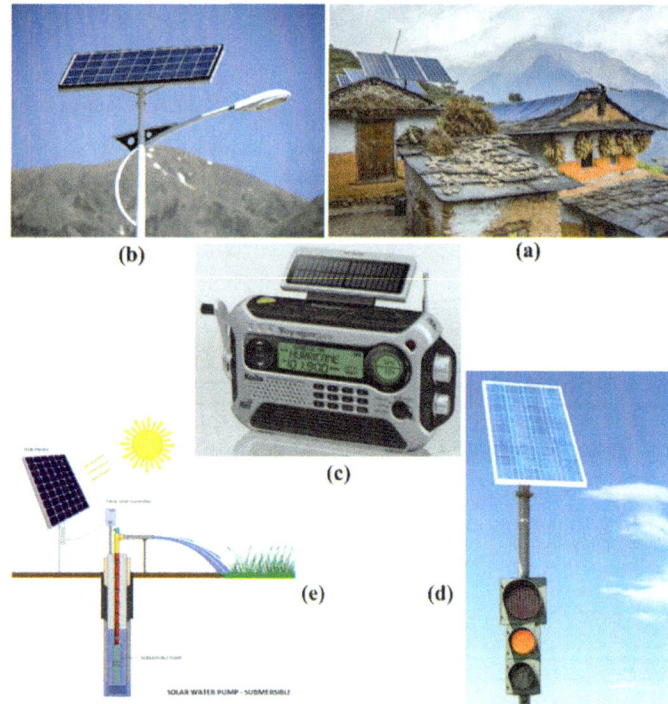

Fig. 2.4 Applications of standalone PV panels in remote areas: **a** rooftops, **b** road lighting, **c** small devices, **d** traffic signalization, and **e** water pumping

2.4 DC–DC Converters

Due to the significant growth in global electric energy demand and the depletion of traditional energy sources such as fossil fuels, PV energy has become the most attractive renewable energy source. Its applications span various sectors, including industry, vehicles, and homes [13–17].

To ensure the effective and reliable utilization of PV energy and to achieve high energy conversion efficiency, DC–DC converters are essential. The implementation of the MPPT mechanism is not feasible without DC–DC converters. They serve as an interface between incompatible components (e.g., PV arrays and loads such as DC charges or batteries) to optimize power transfer from the PV panels to the loads [14, 18].

There are many structures or topologies of DC–DC converters used in PV system applications, including boost, buck, buck-boost, Single-Ended Primary Inductance Converter (SEPIC), forward, flyback, push–pull, and others, as reported in [13, 14, 19–21]. These topologies can be categorized into two classes: isolated and non-isolated DC–DC converters [13–15].

2.4 DC–DC Converters

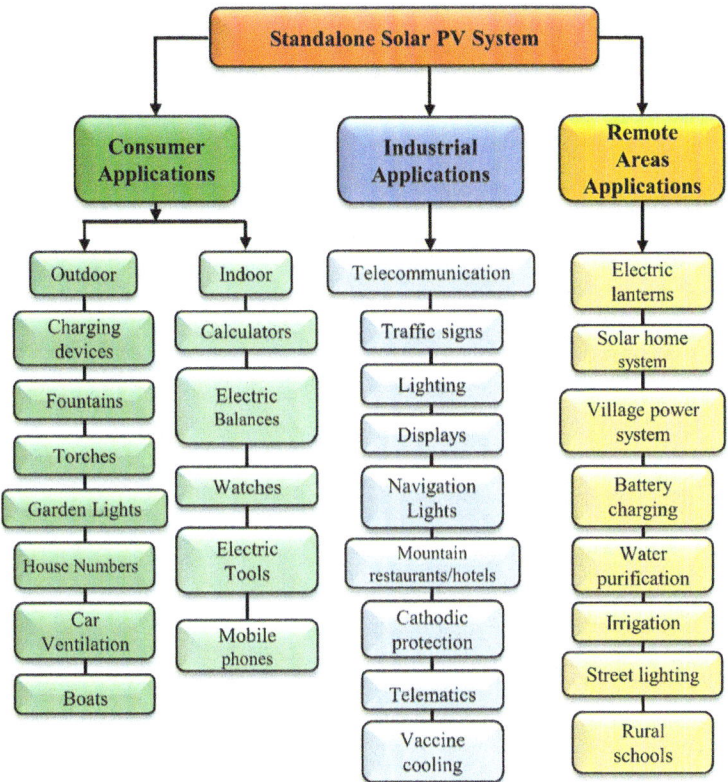

Fig. 2.5 The general applications areas of standalone solar PV systems [4]

This study focuses on non-isolated DC–DC converters due to their simple circuit designs, lower cost, and ability to provide high voltage gain. These characteristics make non-isolated DC–DC converters particularly well-suited for PV applications [20].

Authors in [14, 20] present detailed reviews on non-isolated high-gain DC–DC converters for PV system applications, highlighting several challenges that can arise at high power levels. A new classification has emerged from these reviews, categorizing non-isolated DC–DC converters into four classes:

- Low Gain Low Power Converters (LGLPC): such as the boost converter.
- High Gain Low Power Converters (HGLPC): such as the interleaved boost converter (IBC).
- Low Gain High Power Converters (LGHPC): such as the three-level boost converter (TLBC).
- High Gain High Power Converters (HGHPC): such as the IBC with a coupled inductor (CI).

2.4.1 DC–DC Boost Converter

The DC–DC boost, or step-up converter, shown in Fig. 2.6a, is used in various applications ranging from milliwatts (mW) to gigawatts (GW) of power [21]. Its simple circuit and moderate gain make it the most commonly used non-isolated DC–DC converter in PV system applications [22–27]. Despite these benefits, the boost converter has several drawbacks [33], such as high current ripples and low efficiency. To address these issues, many alternative topologies have been proposed [13, 20, 21, 28].

2.4.1.1 Functioning of the DC–DC Boost Converter

The step-up converter is typically employed to increase an input DC voltage, such as the PV voltage (V_{PV}), to a higher DC voltage according to load demand. This is achieved by controlling the gate (G) of the MOSFET (switch (SW)) with a pulse width modulation (PWM) signal, where the PWM width varies based on the duty ratio (d) ($0 < d < 1$) generated by the controller (e.g., MPPT, PI/PID). As shown in Fig. 2.7, the boost converter operates in two states: enabled (ON) or disabled (OFF).

If the inductor current I_L returns to zero during each switching period, the boost converter operates in discontinuous conduction mode (DCM). Conversely, if is always greater than zero, it operates in continuous conduction mode (CCM). To maintain CCM, the minimum inductor value (L_{min}) is calculated as given in Eq. (2.1) [29, 30]:

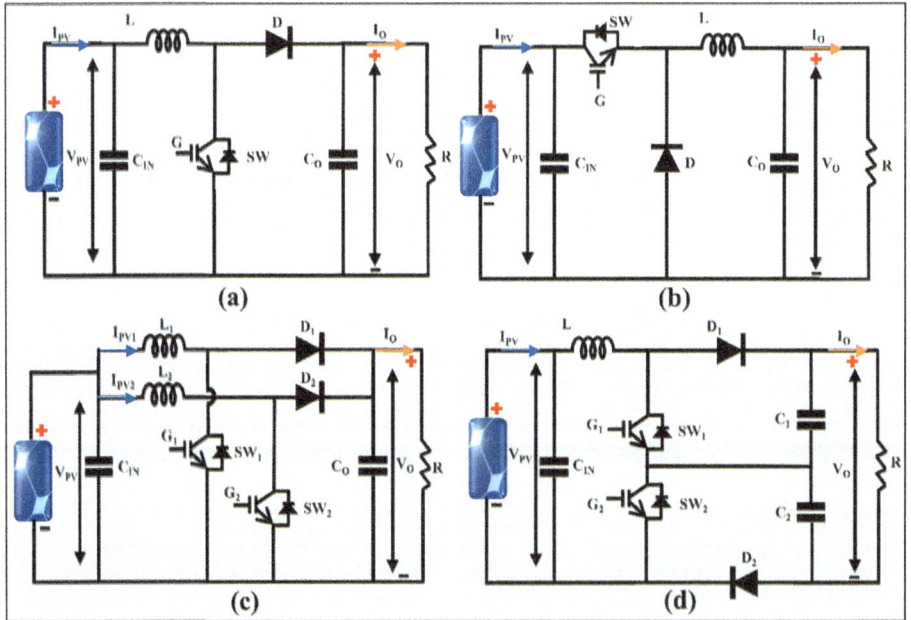

Fig. 2.6 Non-isolated DC–DC converters: **a** boost, **b** buck, **c** interleaved boost, and **d** three-level boost

2.4 DC–DC Converters

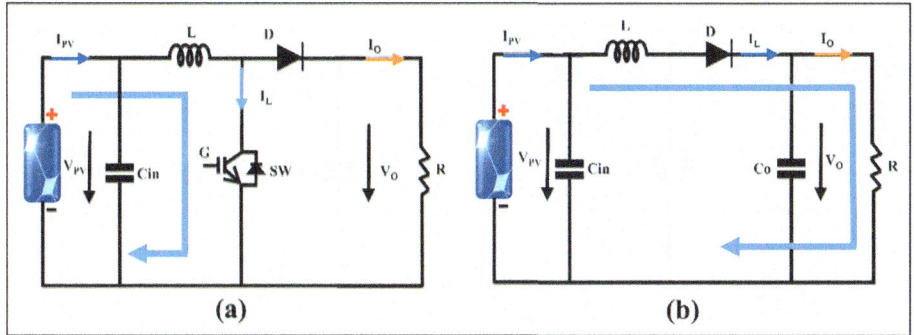

Fig. 2.7 Operating states of the boost converter: **a** ON and **b** OFF

$$L_{min} = \frac{d \times (1-d) \times R}{2 \times F} \quad (2.1)$$

To minimize the ripple in the input and output voltage ($\frac{\Delta V_o}{V_o} = 1\%$) the filter capacitors must be carefully designed and calculated using the following Eqs. (2.2) and (2.3) [29–31]:

$$C_{in} \geq \frac{d}{8 \times F^2 \times \frac{\Delta V_o}{V_o} \times L} = \frac{d}{8 \times F^2 \times 0.01 \times L} \quad (2.2)$$

$$C_o > \frac{d}{\frac{\Delta V_o}{V_o} \times R \times F} = \frac{d}{0.01 \times R \times F} \quad (2.3)$$

The voltage and current conversion ratios of the DC–DC step-up converter are given by Eqs. (2.4) and (2.5).

$$V_{pv} = \frac{V_o}{1-d} \quad (2.4)$$

$$I_{pv} = I_o(1-d) \quad (2.5)$$

By dividing the above equations, the following relationship (Eq. (2.6)) is obtained:

$$R_{pv} = \frac{V_{pv}}{I_{pv}} = \frac{1}{(1-d)^2} \times \frac{V_o}{I_o} = \frac{R}{(1-d)^2} = \frac{R_{Load}}{(1-d)^2} \quad (2.6)$$

where R_{pv} and R_{Load} refer to the resistances seen by the PV array and the load side.
From this relationship, the duty ratio of the step-up converter is given by Eq. (2.7):

$$d = 1 - \sqrt{\frac{R_{pv}}{R_{Load}}} \quad (2.7)$$

2.4.2 DC–DC Buck Converter

The schematic of the DC–DC buck, or step-down, converter illustrated in Fig. 2.6b. Its primary role is to decrease an input DC voltage to match the voltage required by the DC load (e.g., a battery) [4]. In PV system applications, such as off-grid PV pumping systems, standalone PV systems with battery charging stations, and on-grid PV systems, the buck converter is generally employed to adjust the higher PV voltage to the lower DC load voltage and to ensure MPPT tracking [13].

Based on the switch state, the step-down converter operates in two modes as shown in Fig. 2.8. The voltage and current conversion ratios for the buck converter operating in continuous conduction mode (CCM) with ideal components are expressed by Eqs. (2.8) and (2.9) [32]:

$$V_{pv} = V_o \times d \qquad (2.8)$$

$$I_{pv} = \frac{I_o}{d} \qquad (2.9)$$

From these equations, the following relationship (2.10) is derived:

$$R_{pv} = \frac{V_{pv}}{I_{pv}} = (d)^2 \times \frac{V_o}{I_o} = (d)^2 \times R_{Load} \qquad (2.10)$$

Thus, the duty ratio for the buck converter is given by Eq. (2.11) as:

$$d = \sqrt{\frac{R_{pv}}{R_{Load}}} \qquad (2.11)$$

To ensure the buck converter operates in CCM, the inductor value must be greater than the minimum value calculated as:

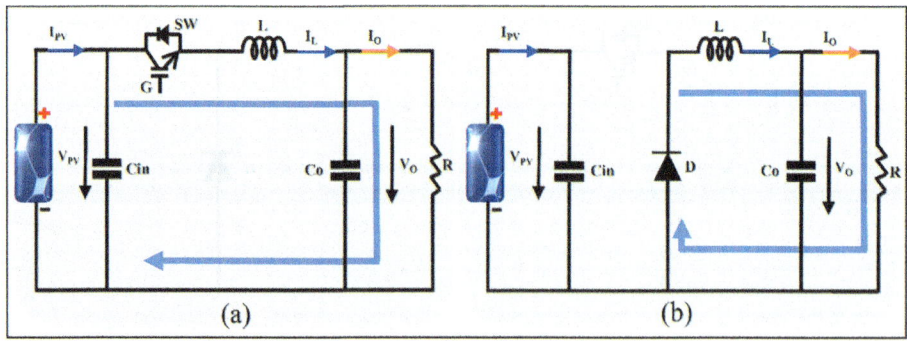

Fig. 2.8 Operating states of the buck converter: **a** ON and **b** OFF

2.4 DC–DC Converters

Table 2.1 Description of IB converter operating modes [38]

Modes	Descriptions
Mode1	The switch SW_1 and diode D_2 are enabled and the switch SW_2 and diode D_1 are disenabled
Mode2	The switch SW_1 and diode D_2 are disenabled and the switch SW_2 and diode D_1 are enabled
Mode3	The switch SW_1 and SW_2 are enabled and the diode D_2 and D_1 are disabled
Mode4	The switch SW_1 and SW_2 are disabled and the diode D_2 and D_1 are enabled

$$L_{min} = \frac{(1-d) \times R}{2 \times F} \quad (2.12)$$

Additionally, to minimize output voltage ripple ($\frac{\Delta V_o}{V_o} = 1\%$), the following equation must be considered:

$$C_o > \frac{1-d}{8 \times F^2 \times \frac{\Delta V_o}{V_o} \times L} = \frac{1-d}{8 \times F^2 \times 0.01 \times L} \quad (2.13)$$

2.4.3 DC–DC Interleaved Boost Converter

As reported in the literature [21, 33], the step-up (boost) converter suffers from various drawbacks that can significantly decrease its gain and efficiency [14]. To address these issues, several architectures have been proposed [14, 19, 21, 34, 35], including the three-level boost (TLB) and interleaved boost (IB) converters.

For the interleaved architecture, numerous studies have shown that the IB is a high-power DC–DC converter interface suitable for renewable energy sources (RES) or energy storage systems (ESS) and the DC bus of grid-connected or off-grid systems [36, 37]. The IB converter is widely used in PV, wind, and fuel cell applications [33, 38, 39] due to its improved overall power efficiency, lower input and output ripple, reduced switching stress, and lower conduction losses due to current splitting [35, 38].

The IB topology, shown in Fig. 2.6c, consists of two inductors, two diodes, an output capacitor filter, and two switches controlled with a 180° phase-shifted PWM signal. The IB operates in four modes, as detailed in Table 2.1 and illustrated in Fig. 2.9.

2.4.4 DC–DC Three-Level Boost Converter

As mentioned previously, the three-level boost converter (TLBC) shown in Fig. 2.6d, is one of the multilevel structures designed to overcome the limitations of the conventional

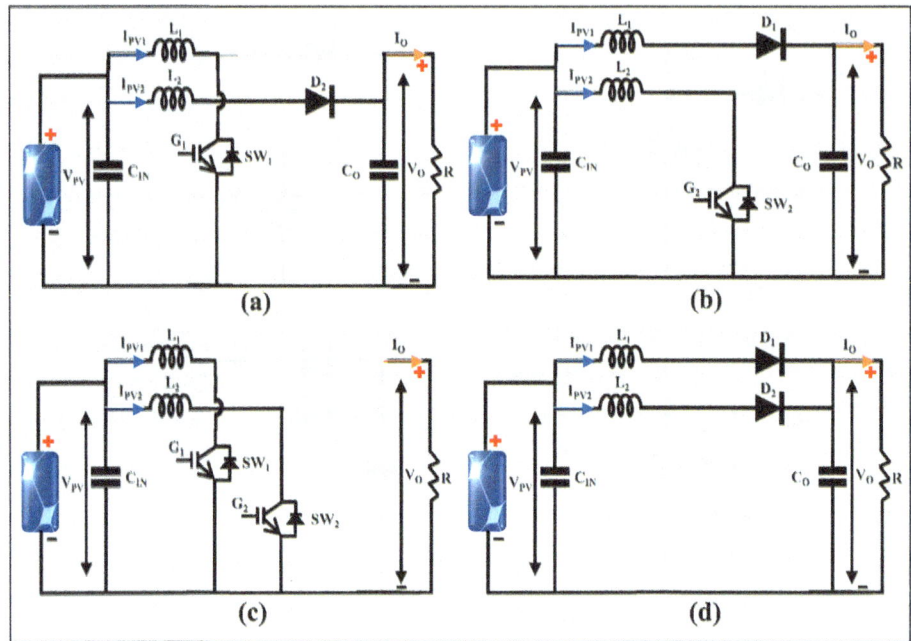

Fig. 2.9 Operating modes of the interleaved boost converter (IB) circuit: **a** mode 1, **b** mode 2, **c** mode 3, and **d** mode 4

boost converter and enhance power conversion efficiency [40]. The TLBC has numerous advantages that make it suitable for integration in high-power conversion applications such as automotive systems [41], AC/DC power factor correction (PFC) applications [42, 43], fuel cell applications [44, 45], photovoltaic systems [46–48], and wind energy applications [49]. However, a significant drawback of the TLBC is the potential for unregulated output capacitor voltages V_{C1} and V_{C2}. This issue can be mitigated using a simple proportional-integral (PI) controller [50–52].

In continuous conduction mode (CCM), the TLBC operates in four possible states depending on the conduction status of switches SW1 and SW2, which can either be both conductive (ON) or non-conductive (OFF), or one ON while the other is OFF, as shown in Fig. 2.10. There are two working conditions for the TLBC based on the comparison between the input voltage (V_{PV}) and half of the output voltage ($0.5V_O$) [40, 53, 54].

- If $V_{PV} \geq 0.5V_O$ meaning the duty ratios of switches SW1 and SW2 are less than or equal to 0.5 ($d_1 \leq 0.5$ and $d_2 \leq 0.5$), the TLBC alternates between operating modes (b), (c), and (d).
- If $V_{PV} < 0.5V_O$, meaning that $d_1 > 0.5$ and $d_2 > 0.5$, the TLBC operates in modes (a), (b), and (d).

2.5 Maximum Power Point Tracking (MPPT) Algorithm

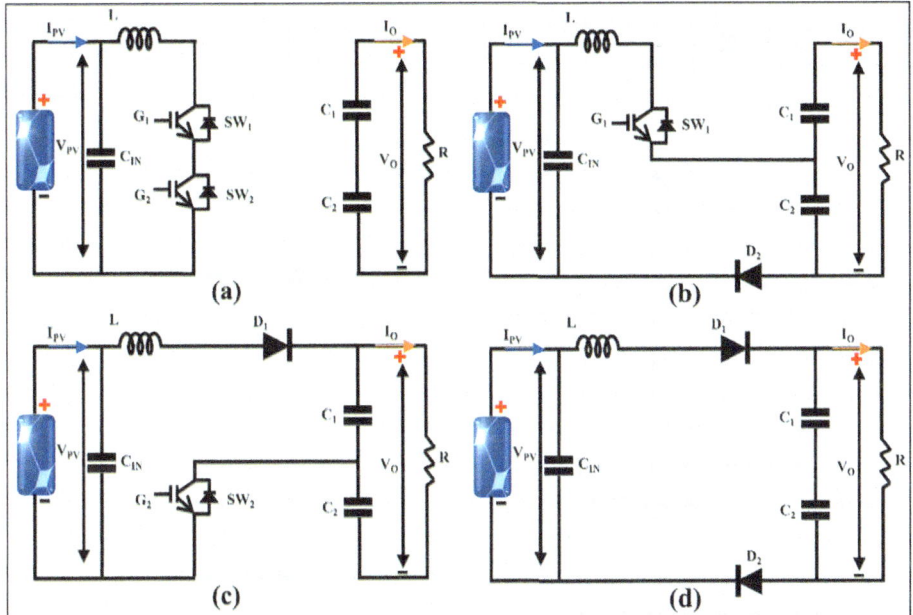

Fig. 2.10 Operating modes of the three-level boost converter (TLBC) circuit: **a** mode 1, **b** mode 2, **c** mode 3, and **d** mode 4

2.5 Maximum Power Point Tracking (MPPT) Algorithm

2.5.1 Importance of the MPPT Algorithm in Solar PV Systems

Despite the advantages of solar photovoltaic energy, several challenges must be addressed to maximize energy yield. The current and power characteristics of a photovoltaic module or array under normal environmental conditions (fixed insolation and temperature) are non-linear and exhibit a unique optimum point, known as the MPP, where the PV array produces maximum power, as shown in Fig. 2.11 [55]. This optimum point varies with atmospheric conditions, particularly solar irradiation and temperature, as well as the electrical characteristics of loads [56, 57]. Therefore, implementing MPPT is essential for enhancing the economic, electrical, and practical performance of PV systems [58, 59].

2.5.2 Essential Specifications of MPPT Algorithm Design

Designing an accurate MPPT algorithm involves considering several factors, such as simplicity and implementation cost, stability, dynamic response, robustness to disturbances,

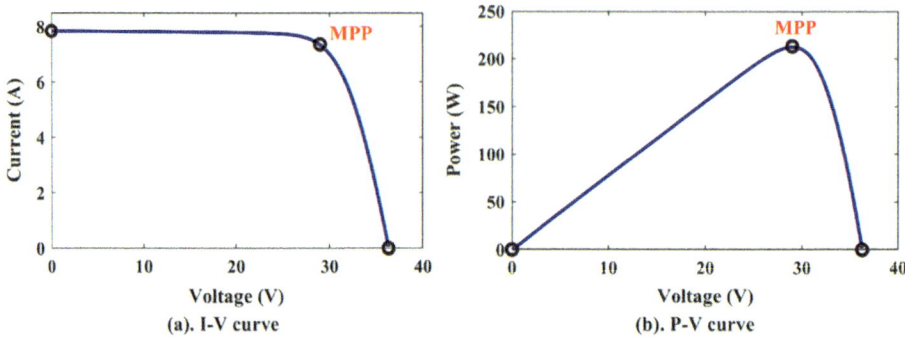

Fig. 2.11 *P–V* and *I-V* characteristics of a photovoltaic array

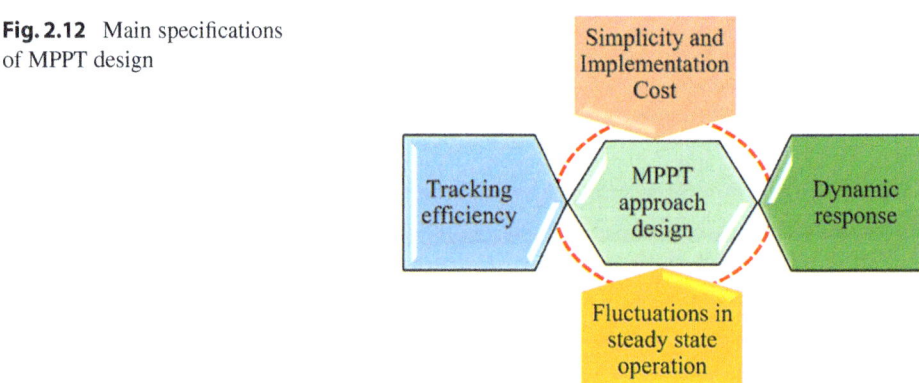

Fig. 2.12 Main specifications of MPPT design

steady-state operation fluctuations, and tracking efficiency, as depicted in Fig. 2.12 [60, 61]. The most critical aspects include [62].

2.5.2.1 Simplicity and Implementation Cost

The complexity and cost of implementing an MPPT algorithm significantly affect its efficiency and accuracy. A good MPPT approach should be simple, robust, and cost-effective.

2.5.2.2 Fluctuations in Steady-State Operation

In steady-state operation, the MPPT algorithm should track the MPP precisely. However, due to the fixed step size perturbation used in the MPPT algorithm, fluctuations are typically observed around the MPP. To minimize these fluctuations, a variable step size can be used instead of a fixed step size.

2.5.2.3 Dynamic Response

An efficient MPPT algorithm should demonstrate rapid tracking speed and a swift response to fluctuating weather conditions. This ensures optimal performance of the photovoltaic system under varying environmental factors.

2.5.2.4 Efficiency of Tracking

Tracking efficiency is a crucial criterion that indicates the effectiveness of an MPPT algorithm in tracking the MPP. It also allows for the comparison of the accuracy of different MPPT methods [62]. The tracking efficiency is calculated by dividing the instantaneous power tracked by the MPPT algorithm ($P_{MPPT}(k)$) by the maximum available power of the PV array ($P_{PVa}(k)$) [63], as shown in Eq. (2.14).

$$\eta_{MPPT} = \frac{P_{MPPT}(k)}{P_{PVa}(k)} \times 100 \qquad (2.14)$$

2.5.3 Description and Classification of MPPT Algorithms in PV Systems

As discussed earlier, the role of the MPPT mechanism in PV systems is indispensable. Consequently, this field has captivated researchers and industrial stakeholders alike [64]. Numerous research studies based on the MPPT approach for PV systems are available in the scientific literature [56–62].

Initially, MPPT methods can be classified based on their tracking control, which can be direct or indirect. The first category, known as duty ratio-based MPPT approaches, falls under direct control, while the second category, known as voltage-based MPPT approaches, falls under indirect control [63]. Additionally, other classifications can be based on various factors such as the capability to track the real MPP under uniform and non-uniform (partial shading) conditions, implementation complexity, cost, and more.

Several authors have categorized MPPT approaches into two groups: conventional and soft computing (advanced) MPPT algorithms [59, 66, 67]. Another classification proposes three groups: conventional or direct, indirect, and soft computing MPPT methods [65]. Alternatively, MPPT techniques can be categorized as offline, online, and hybrid [69]. Furthermore, a classification into four groups has been suggested: conventional, soft computing, smart, and modified MPPT approaches [82]. A new classification based on five groups of MPPT methods includes approaches based on constant parameters, measurement and comparison, trial and error, mathematical calculation, and intelligent prediction [56].

Based on these classifications and a review of various articles, it is concluded that the most appropriate classification of MPPT approaches is into three groups:

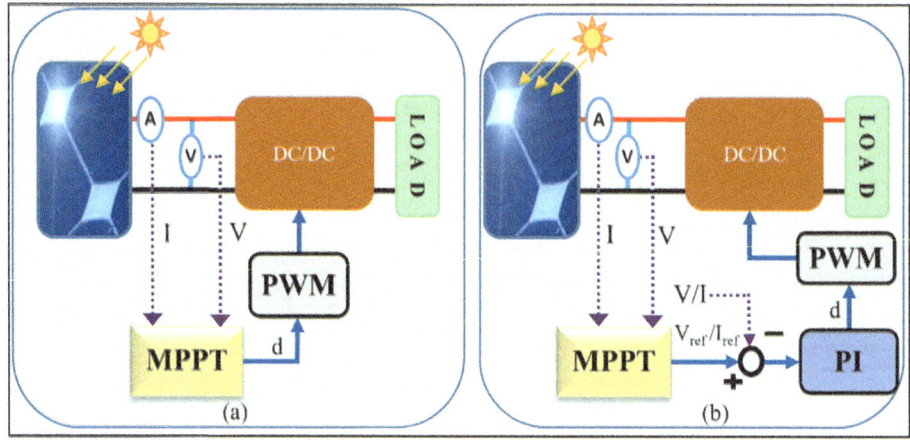

Fig. 2.13 **a** Duty ratio-based and **b** current-/voltage-based MPPT approach implementation

- Conventional MPPT approaches
- Soft Competing (SC) or Artificial Intelligence (AI) MPPT approaches
- Hybrid MPPT approaches.

2.5.3.1 Description of Direct and Indirect MPPT Algorithms

Duty Ratio-Based MPPT Approach

The first implementation structure of the MPPT approach is the duty ratio-based MPPT, also known as direct control, as shown in Fig. 2.13a. In this implementation, the MPPT directly controls the DC–DC converter to operate at the optimal point of the PV array by selecting the optimal duty ratio to change the PWM signal. The main advantages of direct control are simplicity in design, less computation time, and no need for tuning PI controller parameters [65]. Consequently, numerous researchers have adopted this structure in their work [24, 63, 70–77]. However, the duty ratio-based MPPT approach has several drawbacks, such as significant transient fluctuations during load changes and high sensitivity to sudden insolation variations and DC–DC converter topology [68].

Current or Voltage-Based MPPT Approach

Unlike the direct type, the indirect type, also known as the current- or voltage-based MPPT approach, is depicted in Fig. 2.13b. In this implementation, the MPPT block generates a reference current or voltage V_{ref} and I_{ref}, which is compared with the current or voltage of the detected PV module/array. The resulting error is used by a current or voltage controller, generally a PI controller, to generate an adequate duty ratio to extract the available MPP from the PV array [65, 68]. Many papers prefer this type of MPPT approach due to its high tracking efficiency under static and dynamic conditions, short response time, and high tracking speed [25, 78–82].

2.5 Maximum Power Point Tracking (MPPT) Algorithm

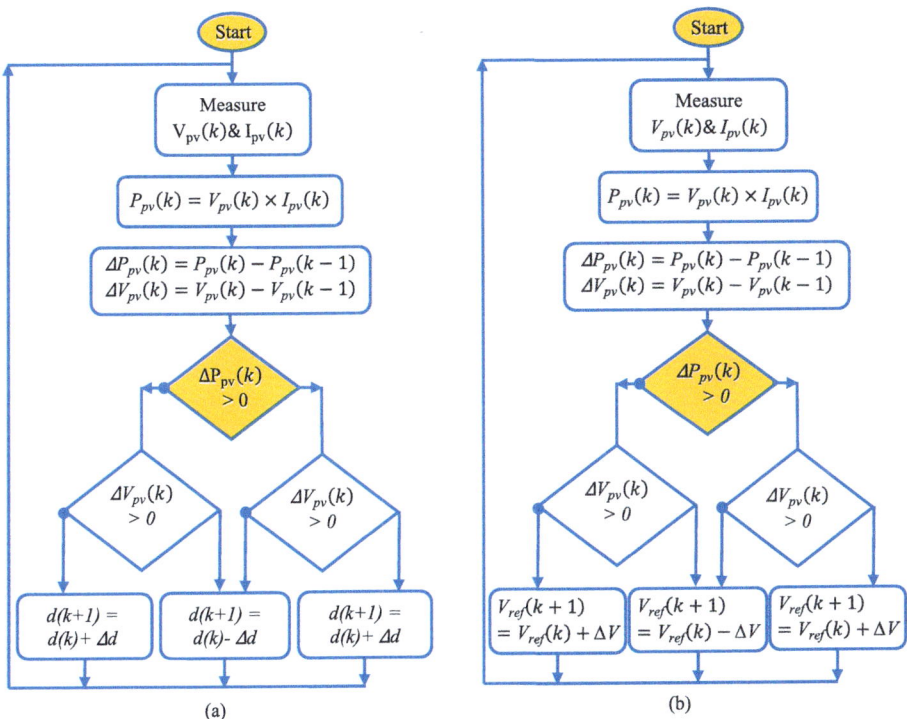

Fig. 2.14 Flowcharts of the P&O MPPT algorithm: **a** Direct control method, **b** Indirect control method

2.5.4 Overview of the Classification of MPPT Algorithms

2.5.4.1 Conventional MPPT Algorithms
Perturbation and Observation (P&O) MPPT Algorithm

Due to its simplicity, ease of use, and low implementation cost, the P&O MPPT algorithm is one of the most popular MPPT algorithms. The principle of P&O is based on perturbing the terminal voltage or current of the PV array and comparing the resulting PV power. If the resulting power increases, the perturbation continues in the same direction; if the power decreases, the direction of perturbation reverses [83].

To implement the P&O MPPT algorithm, there are two methods: direct and indirect control. In the direct control method, the P&O MPPT generates a direct duty cycle for controlling the converter, without requiring a voltage or current controller [68, 84–88]. On the other hand, the indirect control method generates a reference voltage or current, which is compared with the PV voltage or current, and the result is used by the PI controller to generate the appropriate duty ratio [89–92]. The flowcharts of the P&O MPPT algorithm based on both controllers are depicted in Fig. 2.14.

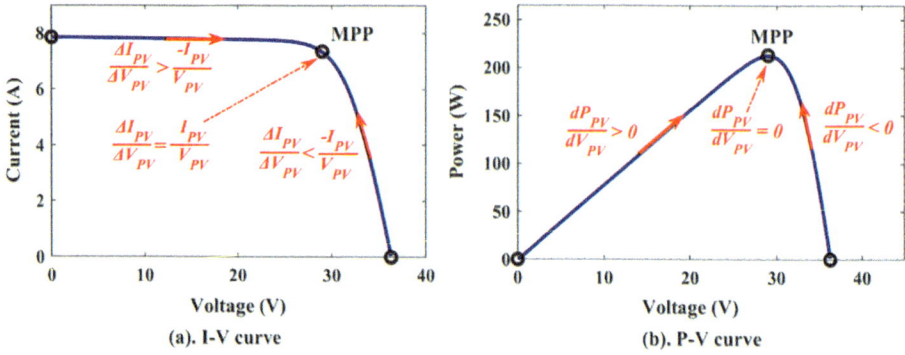

Fig. 2.15 Sing of I_{PV}/V_{PV} and dP_{PV}/dV_{PV} at different position of **a** I–V and **b** P–V curves, respectively

Incremental of Conductance (INC) MPPT Algorithm

Since its introduction by Wasynczuk et al. in 1983 [93], the INC MPPT method has been extensively used and implemented, as revealed by numerous studies in the scientific literature [24, 65, 72, 75, 94–97]. The general idea of the INC MPPT algorithm is based on the differentiation of power with respect to the voltage curves of the PV array array (dP_{pv}/dV_{pv}) as depicted in Fig. 2.15b. At the maximum power point (MPP), $dP_{pv}/dV_{pv} = 0$. To the right of the MPP, $dP_{pv}/dV_{pv} < 0$, and to the left of the MPP, $dP_{pv}/dV_{pv} > 0$ [98]. The working principle of the INC MPPT algorithm to track the MPP can be described mathematically as follows:

$$\frac{dP_{pv}}{dV_{pv}} = I_{pv} + V_{pv}\frac{dI_{pv}}{dV_{pv}} = 0 \text{ which implies}: \frac{dI_{pv}}{dV_{pv}} \cong \frac{\Delta I_{pv}}{\Delta V_{pv}} = -\frac{I_{MPP}}{V_{MPP}} \quad (2.15)$$

The working principle of the INC MPPT algorithm to track the MPP can be described mathematically as follows [30]:

$$\begin{cases} \frac{\Delta I_{pv}}{\Delta V_{pv}} = -\frac{I_{pv}}{V_{pv}} \text{ at MPP} \\ \frac{\Delta I_{pv}}{\Delta V_{pv}} > -\frac{I_{pv}}{V_{pv}} \text{ left of MPP} \\ \frac{\Delta I_{pv}}{\Delta V_{pv}} < -\frac{I_{pv}}{V_{pv}} \text{ right of MPP} \end{cases} \quad (2.16)$$

Therefore, according to the *I-V* curves depicted in Fig. 2.15a, the INC MPPT method tracks the expected MPP when the negative value of the conductance ($-I_{pv}/V_{pv}$) equals the value of the incremental conductance ($\Delta I_{pv}/\Delta V_{pv}$). If the operating point of the PV module or array is far from the MPP, which can be on the right or left side, the INC MPPT method adjusts the operating voltage to bring the system closer to the MPP. Similar to the P&O MPPT scheme, the INC MPPT technique exhibits noticeable fluctuations around

2.5 Maximum Power Point Tracking (MPPT) Algorithm

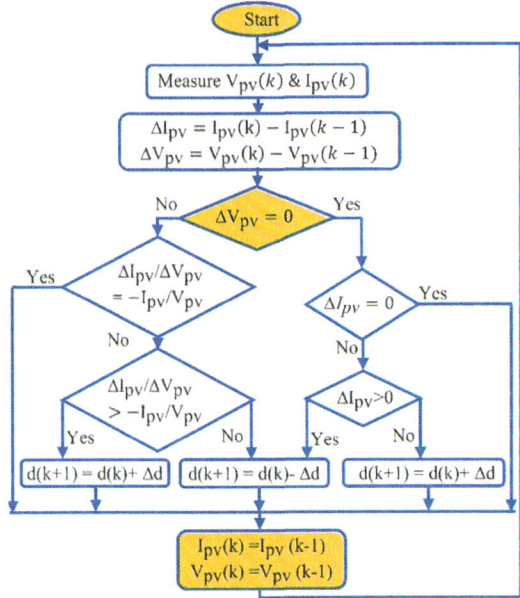

Fig. 2.16 Flowchart of the voltage-based INC MPPT approach

the MPP, and there is a tradeoff between tracking speed and fluctuation. The flowchart of the INC MPPT approach is given in Fig. 2.16.

Hill Climbing (HC) MPPT Algorithm

The HC MPPT algorithm is the simplest of the direct or conventional MPPT methods, which is why it is widely used in many works as [60, 69, 78, 99–102]. Its principle of operation is similar to that of the P&O MPPT algorithm but uses one condition ($\Delta P_{pv} > 0$) instead of two conditions ($\Delta P_{pv} > 0$ and $\Delta V_{pv} > 0$) to locate the MPP, as described in its flowchart shown in Fig. 2.17.

Other MPPT algorithms in this category, derived from conventional methods, include those with adaptable step size, modified variable step size, enhanced, improved, auto-scaling, and advanced MPPT approaches. These advancements are built on traditional MPPT techniques and are extensively covered in the scientific literature [24, 63, 72, 79, 103].

Additionally, there are simpler MPPT techniques such as: Fig. 2.16.

- Fractional Short Circuit Current (FSCC) MPPT method [104]
- Fractional Open Circuit Voltage (FOCV) MPPT method [105]
- Ripple Correlation Control (RCC) MPPT method [106]

Figures 2.18, 2.19, and 2.20 illustrate the flowcharts for the FSCC, FOCV, and RCC MPPT methods, respectively.

Fig. 2.17 Flowchart of hill climbing (HC) MPPT algorithm

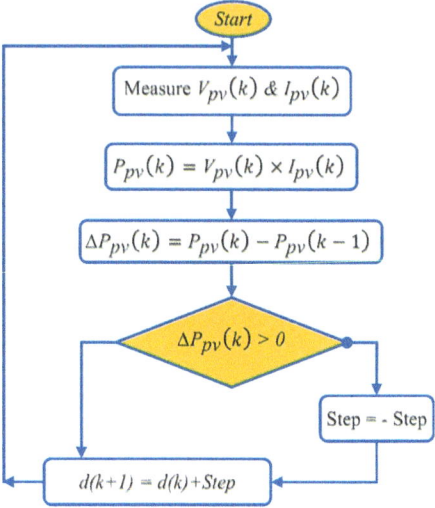

Fig. 2.18 Flowchart of the FSCC MPPT algorithm

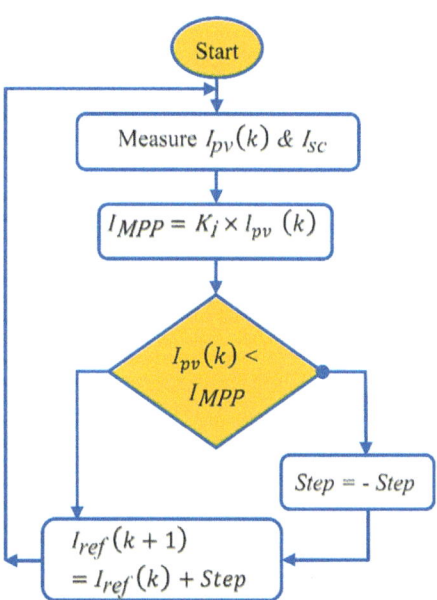

2.5.4.2 Soft Competing (SC) or Artificial Intelligence (AI) MPPT Approaches

With significant advancements in computing power and reduced costs, the integration of intelligent techniques in photovoltaic systems has become increasingly accessible and attractive. SC or AI MPPT methods are widely employed to overcome the limitations of conventional approaches, such as steady-state oscillations, slow tracking speeds, and

2.5 Maximum Power Point Tracking (MPPT) Algorithm

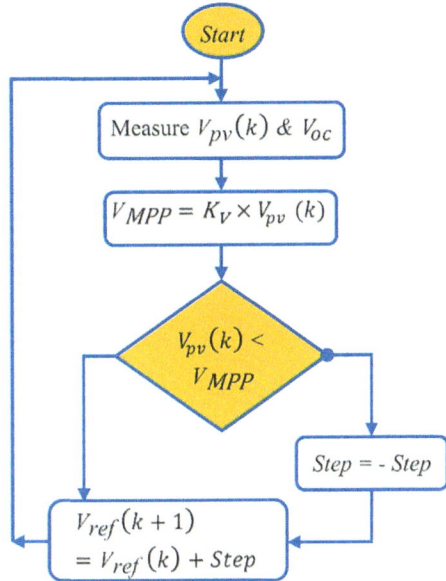

Fig. 2.19 Flowchart of the FOCV MPPT algorithm

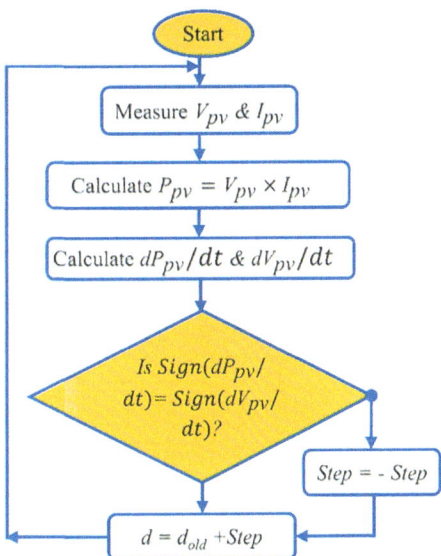

Fig. 2.20 Flowchart of the RCC MPPT algorithm

failure to track the global MPP under partial shading conditions (PSC). These methods offer several advantages, including flexibility, adaptability, and the ability to handle system non-linearities and regulate control functions effectively [107]. Among the most commonly used SC or AI-based MPPT algorithms in PV systems are Fuzzy Logic (FL), Artificial Neural Networks (ANN) and Particle Swarm Optimization (PSO) [67].

2.5.4.3 Fuzzy Logic (FL) MPPT Approach

FLC has recently gained significant attention in PV systems for MPP extraction, primarily due to its strong tracking capabilities and the advantage of not requiring a mathematical model of the system [108]. Consequently, numerous studies have employed FLC as an MPPT approach to efficiently track the MPP [108–114]. However, despite these benefits, the FLC-based MPPT approach has certain drawbacks that can affect its performance and tracking efficiency, such as complexity in implementation, high data storage requirements, extensive computational demands, and its limited effectiveness under PSC [79, 107].

As a result, FLC is often combined with other MPPT methods, such as ANN [115], GA [116], and conventional MPPT algorithms [117, 118] to enhance its overall efficiency and adaptability.

The common structure of an FLC-based MPPT algorithm consists of three main stages: fuzzification, inference, and defuzzification, as illustrated in Fig. 2.21 [108]. In the fuzzification stage, numerical inputs are converted into degrees of linguistic variables, such as NB (Negative Big), PS (Positive Small), NS (Negative Small), PB (Positive Big), and ZE (Zero), using mathematical membership functions. The primary inputs to the FLC-based MPPT algorithm are typically the error (E) and change in error (CE), which are defined by Eqs. (2.17) and (2.18) [91].

$$E(k) = \frac{P_{pv}(k) - P_{pv}(k-1)}{V_{pv}(k) - V_{pv}(k-1)} \, or E(k) = \frac{P_{pv}(k) - P_{pv}(k-1)}{I_{pv}(k) - I_{pv}(k-1)} \tag{2.17}$$

$$CE(k) = E(k) - E(k-1) \tag{2.18}$$

In the inference stage, the FLC MPPT algorithm uses a rule table to compute the output. In the final stage, the defuzzification process converts the linguistic variables back

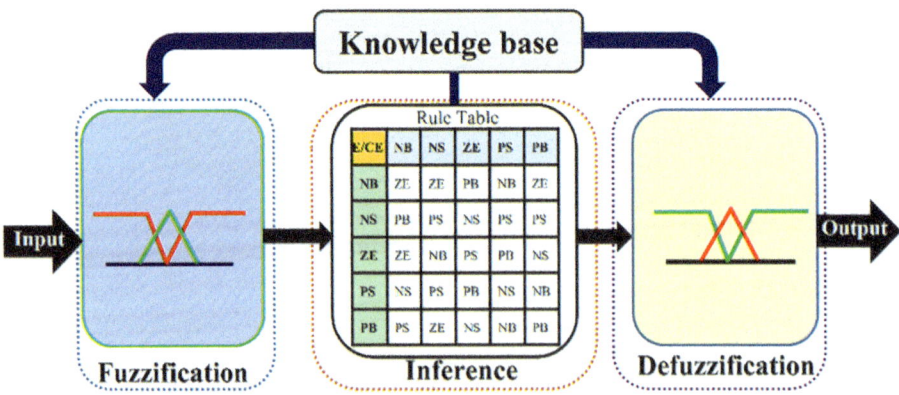

Fig. 2.21 A basic structure of the FLC MPPT approach

2.5 Maximum Power Point Tracking (MPPT) Algorithm

into numerical values, resulting in the output. In this case, the output is the change in the switching duty cycle (ΔD) of the DC–DC converter [119].

2.5.4.4 Artificial Neural Network (ANN) MPPT Approach

In recent years, ANN have been effectively utilized for MPP tracking due to their ability to provide accurate solutions for nonlinear system challenges. The ANN approach is defined by its architecture, consisting of numerous interconnected nodes (artificial neurons). These nodes are organized into three layers: the input layer, hidden layer(s), and output layer, as illustrated in Fig. 2.22 [107]. The input variables for the ANN MPPT approach can include characteristics of the PV array, such as PV current (I_{pv}) and voltage (V_{pv}), atmospheric data such as solar irradiance (G) and ambient temperature (T), or a combination of these. The output(s) typically include reference signals such as MPP current (I_{MPP}), MPP voltage (V_{MPP}), or the duty cycle for controlling the DC–DC converter [82, 120].

Since its initial application for MPP tracking by Hiyama et al [121], numerous studies have explored the use of ANN for MPPT to ensure optimal performance [122, 123]. However, the ANN MPPT technique has some limitations that may reduce its accuracy and effectiveness. One significant drawback is that the training process must be tailored specifically to the PV array and the environmental conditions in which it will operate [124]. Additionally, the ANN must undergo periodic retraining and testing over extended periods (months to years) to ensure reliable tracking of the optimal power point under varying weather conditions [107].

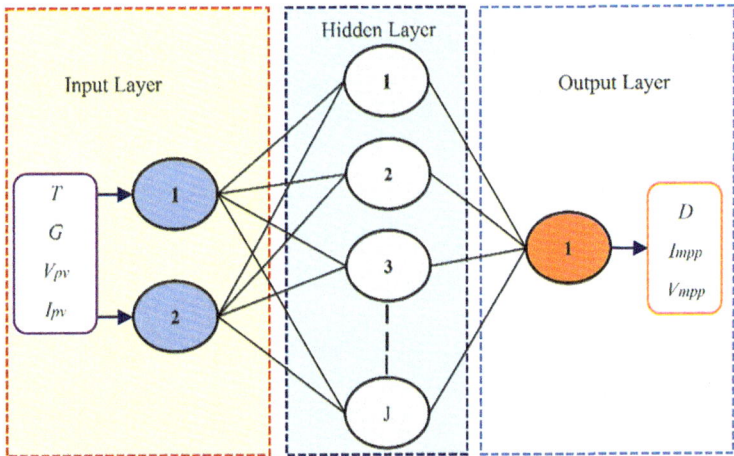

Fig. 2.22 A basic structure of the ANN MPPT approach

2.5.4.5 Particle Swarm Optimization (PSO) MPPT Approach

PSO is an optimization algorithm introduced by Eberhart and Kennedy in 1995, inspired by the social behavior of bird flocks [58]. In PSO, a swarm of agents (known as particles) represents candidate solutions [125]. Each particle moves through the search space with a velocity (v_i^k), which is influenced by the particle's own best-known position (p_{ibest}) and the best-known position of the entire swarm (g_{ibest}) [126], as illustrated in Fig. 2.23. The velocity and position of each particle are updated using the following Eqs. (2.19) and (2.20):

$$v_i^{k+1} = wv_i^k + c_1 r_1 \left(p_{ibest} - x_i^k \right) + c_2 r_2 \left(g_{ibest} - x_i^k \right) \quad (2.19)$$

$$x_i^{k+1} = x_i^k - v_i^{k+1} \quad (2.20)$$

where w represents the inertia weight, c_1 and c_2 are acceleration coefficients; r_1, r_2 are random numbers uniformly distributed between 0 and 1 [126].

In the context of MPP tracking, the particle's position x_i^k represents the duty cycle, and its velocity corresponds to the perturbation. After several iterations, all particles converge toward the duty cycle associated with the MPP. Once the optimal point is reached, the perturbation ceases, and the duty cycle remains constant, ensuring the PV system tracks the MPP [107]. The flowchart of the PSO MPPT algorithm is depicted in Fig. 2.24.

The PSO MPPT approach offers several advantages, including flexibility, robustness, and fast convergence, which have led to its widespread use in PV systems for MPP tracking [127–131]. However, under partial shading conditions (PSC), the PSO algorithm often converges to a local MPP rather than the global MPP (GMPP), limiting its ability to achieve the true optimal power output [58].

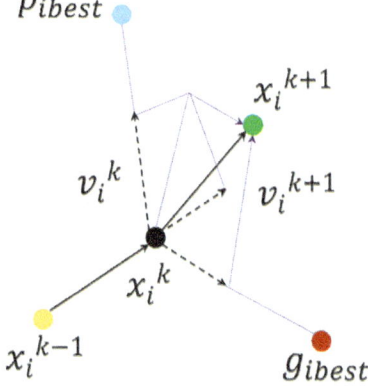

Fig. 2.23 Displacement of particles in the optimization process of the PSO algorithm

2.5 Maximum Power Point Tracking (MPPT) Algorithm

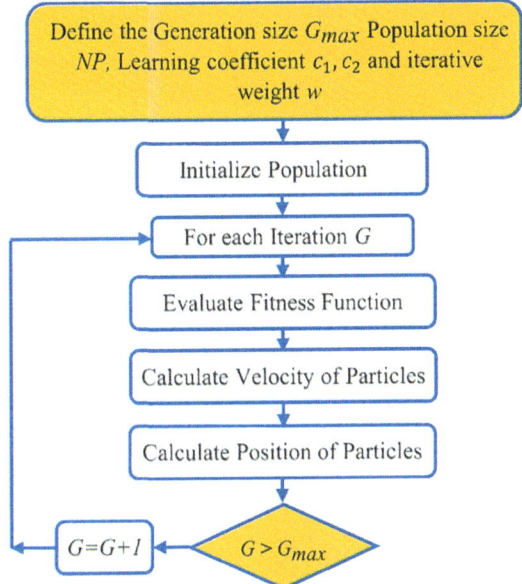

Fig. 2.24 Flowchart of the PSO MPPT algorithm

In addition to the MPPT techniques mentioned above, other SC and AI methods have been extensively applied to ensure high accuracy in MPP tracking, as highlighted in [132]. Examples include:

- Genetic Algorithm (GA) MPPT approach [133]
- Cuckoo Search (CS) MPPT method [134]
- Artificial Bee Colony (ABC) MPPT method [135]
- Ant Colony Optimization (ACO) MPPT method [136]
- Grey Wolf Optimization (GWO) MPPT method [137]
- Firefly Algorithm (FFA) MPPT method [138]
- Fireworks Algorithm (FWA) MPPT method [139].

2.5.4.6 Hybrid MPPT Approaches

Hybrid MPPT approaches, discussed in this subsection, are created by combining two MPPT techniques from the same group or different groups, such as conventional with AI (C-AI) or AI with AI (AI-AI). These hybrid techniques have been proposed because neither conventional nor AI-based MPPT methods alone can consistently deliver optimal tracking performance under challenging climatic conditions (e.g., rapid changes, multiple power peaks in partial shading conditions). Hybrid methods offer key benefits such as versatility, robustness, and fault tolerance [140]. Among hybrid approaches, those combining conventional and AI techniques (C-AI) are more commonly used because they merge the simplicity of conventional methods with the robustness of AI techniques. Numerous hybrid techniques based on C-AI MPPT approaches have been developed, such as:

- PSO with P&O (PSO-P&O) [141, 142]
- ANN with P&O (ANN-P&O) [143, 144]
- FL with P&O (FL-P&O) [88]
- Grey Wolf with P&O (GWO-P&O) [85]
- FL with INC (FL-INC) [145, 146]
- FFA with INC (FFA-INC) [147]
- ANN with HC (ANN-HC) [148, 149].

The second category of hybrid MPPT approaches combines two AI techniques (AI-AI) to improve tracking efficiency and overcome drawbacks, such as failure to track the global GMPP under complex partial shading conditions and long convergence times. Various AI-AI hybrid techniques have been proposed, such as:

- FL with ANN (FL-ANN) [150, 151]
- GWO with FL (GWO-FL) [152]
- ANN with ACO (ANN-ACO) [153]
- PSO with FWA (PSO-FWA) [154]
- GA with ACO (GA-ACO) [140, 155].

For additional hybrid MPPT methods, the reviews in [93, 132, 156] provide comprehensive overviews of the most relevant hybrid MPPT algorithms for PV systems. Despite their advantages, hybrid MPPT approaches face challenges, with implementation complexity being one of the most significant [156].

2.6 Battery Technologies

The global demand for electrical energy has surged significantly in recent years, which can be attributed to the growing installations of renewable energy systems. As a result, the need for energy storage systems to capture excess energy and make it available for later use has become essential. Energy storage systems are a critical component in enhancing the integration and penetration of renewable energy sources [157]. Various energy storage technologies have been proposed, including batteries, capacitors and supercapacitors, pumped hydro, flywheels, compressed air storage, and others, as reported in [158, 159].

In PV systems, particularly in standalone (off-grid) configurations, batteries are the most commonly used energy storage solution. Their role in storing surplus energy

2.6 Battery Technologies

improves system performance and reliability [18, 145]. Batteries are essential in addressing challenges such as power intermittency, bidirectional energy flow requirements, and peak demand management [157].

2.6.1 Overview of Battery Technologies

A battery, or accumulator, is a device capable of converting electrical energy into storable chemical energy and back again, using an electrochemical or chemical-electrical conversion process within an electrochemical cell [160]. A battery typically consists of several electrochemical cells arranged in series or parallel configurations to provide the desired voltage and current [161]. Each electrochemical cell is the core of the battery's operation, comprising three key components: two electrodes (anode and cathode) and an electrolyte [162]. The performance of a battery is significantly influenced by three factors: nominal capacity, state of charge (SOC) limits, and charge rate [163].

Battery technology has advanced considerably, and a variety of battery types have been developed for use in power system applications. These include lead-acid, sodium-sulfur (NaS), lithium-ion (Li-ion), nickel-cadmium (NiCd), sodium-nickel chloride (NaNiCl2) (also known as Zero Emission Battery Research Activities or ZEBRA), zinc-bromide (Zn-Br), polysulfide bromine (PSB), ultra-batteries, and metal-air batteries [157, 161]. Among these, lead-acid, NiCd, and Li-ion batteries are the most commonly deployed in PV systems, particularly in standalone (off-grid) PV applications [160, 164].

2.6.1.1 Lead-Acid Battery

The lead-acid battery, introduced in 1859 by French physicist Gaston Planté, is the oldest form of electrochemical battery technology [157]. During the charging process, it consists of a positive electrode made of lead dioxide (PbO_2), a negative electrode composed of sponge lead (Pb), and an electrolyte of sulfuric acid in water [157, 165]. During discharge, a reaction between the lead dioxide, lead, and sulfuric acid produces lead sulfate ($PbSO_4$), as illustrated in the basic electrochemical reaction in Eq. (2.21) [157, 161, 165, 166]:

$$Pb + PbO_2 + 2HSO_4 \rightleftharpoons 2PbSO_4 + 2H_2O \tag{2.21}$$

There are two main types of lead-acid batteries: flooded and valve-regulated (or sealed) types [161, 167]. Due to its numerous advantages, such as low cost, maturity, and widespread availability, lead-acid technology has been widely used in various applications, including PV systems [71, 163, 168–170]. However, lead-acid batteries have certain drawbacks, such as high maintenance requirements, lower efficiency, short cycle life, slow charging, relatively short lifespan, and low energy and power density [157, 174].

2.6.1.2 Nickel–Cadmium (Ni–Cd)

The nickel-cadmium (Ni–Cd) battery is another mature technology, offering significant improvements in energy density and reduced maintenance compared to lead-acid batteries. Its robustness, long lifespan, low cycling degradation, and reliability make it suitable for grid-independent PV systems [169, 171]. However, its high cost, memory effect, and negative environmental impact are notable disadvantages [167].

In a Ni–Cd battery, the negative electrode (cadmium) is metallic cadmium (Cd) and cadmium hydroxide (Cd(OH)$_2$), while the positive electrode (nickel) is nickel hydroxide Ni(OH)$_2$. The electrolyte is an aqueous solution of potassium hydroxide (KOH) [157, 172]. The electrochemical reaction in a Ni–Cd battery is described by Eq. (2.22):

$$2NiO(OH) + Cd + 2HO_2 \rightleftharpoons 2NiO(OH)_2 + Cd(OH)_2 \qquad (2.22)$$

2.6.1.3 Lithium-Ion (Li-Ion)

Lithium-ion (li-ion) battery operation involves the migration of lithium ions from the positive electrode (anode) to the negative electrode (cathode) through an electrolyte consisting of lithium salts dissolved in organic carbonates during charging [167]. The process is reversed during discharge. The cathode in Li-ion batteries is typically composed of lithium metal oxides such as LiNiO$_2$, LiMO$_2$ or LiCoO$_2$, while the anode is made of graphitic carbon [157]. The electrochemical reactions governing Li-ion battery operation are shown in Eqs. (2.23) and (2.24) [157, 173]:

$$Li_{1-x}MO_2 + xLi^+ + xe^- \rightleftharpoons LiMO_2 \qquad (2.23)$$

$$C + xLi^+ + xe^- \rightleftharpoons Li_xC \qquad (2.24)$$

Li-ion batteries are widely used due to their high energy density and the potential to achieve nearly 100% energy storage efficiency. In 2015, Li-ion batteries accounted for 85.6% of deployed energy storage systems globally [167, 169, 174]. Despite these advantages, Li-ion technology has high upfront costs and requires complex charge management systems due to safety concerns [163, 175].

2.7 Summary

This chapter provides a comprehensive literature review of the key components of a standalone solar PV system, focusing on MPPT algorithms, DC–DC converters, and batteries. The review begins by highlighting solar energy applications and the various types of solar PV systems. Additionally, the most commonly used DC–DC converters and battery technologies are discussed. Furthermore, the widely adopted MPPT techniques for achieving

MPP tracking in solar PV systems are reviewed, categorized into three groups: conventional methods, soft computing or artificial intelligence techniques, and hybrid MPPT algorithms.

Conventional MPPT methods, such as P&O and INC, are characterized by their simplicity and low-cost implementation. However, they exhibit poor tracking performance under rapidly changing environmental conditions (e.g., temperature and solar irradiance) and suffer from a trade-off between steady-state oscillations and tracking speed. On the other hand, SC or AI-based MPPT techniques address some of these limitations by offering improved performance. However, their complexity in hardware implementation, the need for periodic tuning, and their tendency to fail in complex PSC are significant drawbacks.

To overcome the limitations of conventional and AI-based techniques, hybrid MPPT methods have been introduced. While these approaches offer enhanced robustness and tracking accuracy, they come with increased hardware complexity and longer convergence times, which may reduce their overall tracking efficiency.

To address these challenges, this thesis proposes novel and improved MPPT approaches that aim to blend the simplicity of conventional methods with the effectiveness of advanced techniques. These proposed methods offer an optimal balance, providing both simplicity and high tracking performance.

References

1. C. Lupangu, R.C. Bansal, A review of technical issues on the development of solar photovoltaic systems. Renew. Sustain. Energy Rev. **73** (2017). https://doi.org/10.1016/j.rser.2017.02.003.
2. International Energy Agency, CO_2 emissions in 2023 (2023). https://www.iea.org/reports/co2-emissions-in-2023
3. M.B. Hayat, D. Ali, K.C. Monyake, L. Alagha, N. Ahmed, Solar energy—A look into power generation, challenges, and a solar-powered future. Int. J. Energy Res. **43**(3), 1049–1067 (2019)
4. A. Luque, S. Hegedus, *Handbook of Photovoltaic Science and Engineering* (2011). https://doi.org/10.1002/9780470974704
5. SolarPower Europe, Global market outlook for solar power 2024–2028 (2024). https://www.solarpowereurope.org/insights/outlooks/global-market-outlook-for-solar-power-2024-2028#download
6. IRENA, Renewable capacity statistics 2024 (2024). www.irena.org
7. S. Goel, R. Sharma, Performance evaluation of stand alone, grid connected and hybrid renewable energy systems for rural application: a comparative review. Renew. Sustain. Energy Rev. **78**, 1378–1389 (2017). https://doi.org/10.1016/J.RSER.2017.05.200
8. IRENA, Off-grid renewable energy solutions to expand electricity access : an opportunity not to be missed (Abu Dhabi, 2019). https://irena.org/publications/2019/Jan/Off-grid-renewable-energy-solutions-to-expand-electricity-to-access-An-opportunity-not-to-be-missed

9. M. Obi, R. Bass, Trends and challenges of grid-connected photovoltaic systems—a review. Renew. Sustain. Energy Rev. **58**, 1082–1094 (Elsevier Ltd, 2016). https://doi.org/10.1016/j.rser.2015.12.289
10. WHO, IEA, IRENA, UNSD, World Bank, Tracking SDG 7: the energy progress report. https://trackingsdg7.esmap.org/
11. IRENA, Tracking SDG 7: the energy progress report (Abu Dhabi, 2021). https://www.irena.org/publications/2021/Jun/Tracking-SDG-7-2021
12. IRENA, Future of solar photovoltaic: deployment, investment, technology, grid integration and socio-economic aspects (A global energy transformation paper) (International Renewable Energy Agency, Abu Dhabi, 2019)
13. V.G.R. Kummara et al. A comprehensive review of DC–DC converter topologies and modulation strategies with recent advances in solar photovoltaic systems. Electronics (Switzerland) **9**(1) (2020). https://doi.org/10.3390/electronics9010031
14. B.S. Revathi, M. Prabhakar, Non isolated high gain DC–DC converter topologies for PV applications—a comprehensive review. Renew. Sustain. Energy Rev. **66** (2016). https://doi.org/10.1016/j.rser.2016.08.057
15. R.R. de Melo, F.L. Tofoli, S. Daher, F.L.M. Antunes, Interleaved bidirectional DC–DC converter for electric vehicle applications based on multiple energy storage devices. Electr. Eng. **102**(4) (2020). https://doi.org/10.1007/s00202-020-01009-3
16. A. Allouhi et al., PV water pumping systems for domestic uses in remote areas: sizing process, simulation and economic evaluation. Renew. Energy **132**, 798–812 (2019). https://doi.org/10.1016/j.renene.2018.08.019
17. K.J. Reddy, S. Natarajan, Energy sources and multi-input DC–DC converters used in hybrid electric vehicle applications—a review. Int. J. Hydrogen Energy **43**(36), 17387–17408 (Elsevier Ltd., 2018). https://doi.org/10.1016/j.ijhydene.2018.07.076
18. H. Louie, *Off-Grid Electrical Systems in Developing Countries* (2018). https://doi.org/10.1007/978-3-319-91890-7
19. S. Sivakumar, M.J. Sathik, P.S. Manoj, G. Sundararajan, An assessment on performance of DC–DC converters for renewable energy applications. Renew. Sustain. Energy Rev. **58**, 1475–1485 (Elsevier Ltd, 2016). https://doi.org/10.1016/j.rser.2015.12.057.
20. M.H. Taghvaee, M.A.M. Radzi, S.M. Moosavain, H. Hizam, M.H. Marhaban, A current and future study on non-isolated DC–DC converters for photovoltaic applications. Renew. Sustain. Energy Rev. **17** (2013). https://doi.org/10.1016/j.rser.2012.09.023
21. M. Forouzesh, Y.P. Siwakoti, S.A. Gorji, F. Blaabjerg, B. Lehman, Step-up DC–DC converters: a comprehensive review of voltage-boosting techniques, topologies, and applications. IEEE Trans. Power Electron. **32**(12) (2017). https://doi.org/10.1109/TPEL.2017.2652318
22. S. Jana, N. Kumar, R. Mishra, D. Sen, T.K. Saha, Development and implementation of modified MPPT algorithm for boost converter-based PV system under input and load deviation. Int. Trans. Electr. Energy Syst. **30**(2) (2020). https://doi.org/10.1002/2050-7038.12190
23. R. Ayop, C.W. Tan, Design of boost converter based on maximum power point resistance for photovoltaic applications. Sol. Energy **160**, 322–335 (2018). https://doi.org/10.1016/j.solener.2017.12.016
24. S. Motahhir, A. Chalh, A. El Ghzizal, A. Derouich, Development of a low-cost PV system using an improved INC algorithm and a PV panel Proteus model. J. Clean. Prod. **204** (2018). https://doi.org/10.1016/j.jclepro.2018.08.246
25. A. Kihal, F. Krim, A. Laib, B. Talbi, H. Afghoul, An improved MPPT scheme employing adaptive integral derivative sliding mode control for photovoltaic systems under fast irradiation changes. ISA Trans. **87** (2019). https://doi.org/10.1016/j.isatra.2018.11.020

26. P. Shaw, Modelling and analysis of an analogue MPPT-based PV battery charging system utilising DC–DC boost converter. IET Renew. Power Gener. **13**(11) (2019). https://doi.org/10.1049/iet-rpg.2018.6273
27. W. Xiao, N. Ozog, W.G. Dunford, Topology study of photovoltaic interface for maximum power point tracking. IEEE Trans. Ind. Electron. **54**(3) (2007). https://doi.org/10.1109/TIE.2007.894732
28. S. Saravanan, N.R. Babu, A modified high step-up non-isolated DC–DC converter for PV application. J. Appl. Res. Technol. **15**(3) (2017). https://doi.org/10.1016/j.jart.2016.12.008
29. M.H. Rashid, *Power Electronics Handbook* (2007). https://doi.org/10.1016/B978-0-12-088479-7.X5018-4
30. S. Motahhir, Contribution to the optimization of energy withdrawn from a PV panel using an embedded system. SSRN Electron. J. (2021). https://doi.org/10.2139/ssrn.3804081
31. N. Mohan, T.M. Undeland, W.P. Robbins, *Power Electronics: Converters, Applications, and Design* (John Wiley & Sons, 2003)
32. D. Choudhary, A.R. Saxena, DC–DC buck-converter for MPPT of PV system. Int. J. Emerg. Technol. Adv. Eng. **4**(7) (2014)
33. D.Y. Jung, Y.H. Ji, S.H. Park, Y.C. Jung, C.Y. Won, Interleaved soft-switching boost converter for photovoltaic power-generation system. IEEE Trans. Power Electron. **26**(4) (2011). https://doi.org/10.1109/TPEL.2010.2090948
34. W. Li, X. He, A family of interleaved DC–DC converters deduced from a basic cell with winding-cross-coupled inductors (WCCIs) for high step-up or step-down conversions. IEEE Trans. Power Electron. **23**(4) (2008). https://doi.org/10.1109/TPEL.2008.925204
35. P. Abishri, S. Umashankar, R. Sudha, Review of coupled two and three phase interleaved boost converter (IBC) and investigation of four phase ibc for renewable application. Int. J. Renew. Energy Res. **6**(2) (2016)
36. A. Khosroshahi, M. Abapour, M. Sabahi, Reliability evaluation of conventional and interleaved DC–DC boost converters. IEEE Trans. Power Electron. **30**(10) (2015). https://doi.org/10.1109/TPEL.2014.2380829
37. Y. Gu, D. Zhang, Interleaved boost converter with ripple cancellation network. IEEE Trans. Power Electron. **28**(8) (2013). https://doi.org/10.1109/TPEL.2012.2228505
38. K.S. Faraj, Design and implementation of high step-up interleaved boost converter based on photovoltaic system application. University of Technology, 2019
39. F. Khoucha, A. Benrabah, O. Herizi, A. Kheloui, M.E.H. Benbouzid, An improved MPPT interleaved boost converter for solar electric vehicle application, in *International Conference on Power Engineering, Energy and Electrical Drives* (2013). https://doi.org/10.1109/PowerEng.2013.6635760
40. D. Oulad-Abbou, S. Doubabi, A. Rachid, Voltage balance control analysis of three-level boost DC–DC converters: theoretical analysis and DSP-based real time implementation. Energies **11**(11) (2018). https://doi.org/10.3390/en11113073
41. S. Dusmez, A. Hasanzadeh, A. Khaligh, Comparative analysis of bidirectional three-level DC–DC converter for automotive applications. IEEE Trans. Ind. Electron. **62**(5) (2015). https://doi.org/10.1109/TIE.2014.2336605
42. H.C. Chen, J.Y. Liao, Modified interleaved current sensorless control for three-level boost PFC converter with considering voltage imbalance and zero-crossing current distortion. IEEE Trans. Ind. Electron. **62**(11) (2015). https://doi.org/10.1109/TIE.2015.2435695
43. B. Mahdavikhah, A. Prodić, Low-volume PFC rectifier based on nonsymmetric multilevel boost converter. IEEE Trans. Power Electron. **30**(3) (2015). https://doi.org/10.1109/TPEL.2014.2317723

44. Y.R. De Novaes, A. Rufer, I. Barbi, A new quadratic, three-level, DC/DC converter suitable for fuel cell applications, in *Fourth Power Conversion Conference-NAGOYA, PCC-NAGOYA 2007–Conference Proceedings* (2007). https://doi.org/10.1109/PCCON.2007.373028
45. A. Shahin, M. Hinaje, J.P. Martin, S. Pierfederici, S. Rael, B. Davat, High voltage ratio DC–DC converter for fuel-cell applications. IEEE Trans. Ind. Electron. **57**(12) (2010) https://doi.org/10.1109/TIE.2010.2045996
46. J.M. Kwon, B.H. Kwon, K.H. Nam, Three-phase photovoltaic system with three-level boosting MPPT control. IEEE Trans. Power Electron. **23**(5) (2008). https://doi.org/10.1109/TPEL.2008.2001906
47. C. Balakishan, N. Sandeep, M.V. Aware, P. Bauer, Design and implementation of three-level DC–DC converter with golden section search based MPPT for the photovoltaic applications. Adv. Power Electron. **2015** (2015). https://doi.org/10.1155/2015/587197
48. J. Rajesh, K.S. Nisha, A.K. Bonala, S.R. Sandepudi, Predictive control of three level boost converter interfaced SPV system for bi-polar DC micro grid, in *Proceedings of 2019 3rd IEEE International Conference on Electrical, Computer and Communication Technologies, ICECCT 2019* (2019). https://doi.org/10.1109/ICECCT.2019.8869216
49. C.H. Tran, F. Nollet, N. Essounbouli, A. Hamzaoui, Maximum power point tracking techniques for wind energy systems using three levels boost converter, in *IOP Conference Series: Earth and Environmental Science* (2018). https://doi.org/10.1088/1755-1315/154/1/012016
50. M. Samadi, S.M. Rakhtala, Reducing cost and size in photovoltaic systems using three-level boost converter based on fuzzy logic controller. Iran. J. Sci. Technol. Trans. Electr. Eng. **43** (2019). https://doi.org/10.1007/s40998-018-0145-6
51. A.K. Mishra, B. Singh, An efficient control scheme of self-reliant solar-powered water pumping system using a three-level DC–DC converter. IEEE J. Emerg. Sel. Top. Power Electron. **8**(4) (2020). https://doi.org/10.1109/JESTPE.2019.2943203
52. A. Chellakhi, S. El Beid, Y. Abouelmahjoub, Y. Mchaouar, An efficient implementation of three-level boost converter with capacitor voltage balancing for an advanced MPPT approach in PV Systems. e-Prime Adv. Electr. Eng. Electron. Energy **9**(April), 100688 (2024). https://doi.org/10.1016/j.prime.2024.100688
53. Y. Zhao, T. Xu, Y. Zhao, S. Hu, Design of double loop controller for three-level boost converter. https://doi.org/10.1088/1757-899X/768/6/062059
54. A. Nouri, I. Salhi, E. Elwarraki, S. El Beid, N. Essounbouli, DSP-based implementation of a self-tuning fuzzy controller for three-level boost converter. Electr. Power Syst. Res. **146** (2017). https://doi.org/10.1016/j.epsr.2017.01.036
55. M. Balato, L. Costanzo, M. Vitelli, DMPPT PV system: modeling and control techniques, in *Advances in Renewable Energies and Power Technologies*, vol. 1 (2018). https://doi.org/10.1016/B978-0-12-812959-3.00005-8
56. N. Karami, N. Moubayed, R. Outbib, General review and classification of different MPPT techniques. Renew. Sustain. Energy Rev. **68** (2017). https://doi.org/10.1016/j.rser.2016.09.132
57. M.A. Husain, A. Tariq, S. Hameed, M.S.B. Arif, A. Jain, Comparative assessment of maximum power point tracking procedures for photovoltaic systems. Green Energy Environ. **2**(1), 5–17 (KeAi Publishing Communications Ltd., 2017). https://doi.org/10.1016/j.gee.2016.11.001
58. L. Liu, X. Meng, C. Liu, A review of maximum power point tracking methods of PV power system at uniform and partial shading. Renew. Sustain. Energy Rev. **53** (2016). https://doi.org/10.1016/j.rser.2015.09.065
59. M.F.N. Tajuddin, M.S. Arif, S.M. Ayob, Z. Salam, Perturbative methods for maximum power point tracking (MPPT) of photovoltaic (PV) systems: a review. Int. J. Energy Res. **39**(9) (2015) https://doi.org/10.1002/er.3289

60. W. Xiao, W.G. Dunford, A modified adaptive hill climbing MPPT method for photovoltaic power systems, in *PESC Record—IEEE Annual Power Electronics Specialists Conference* (2004). https://doi.org/10.1109/PESC.2004.1355417
61. C. Cabal, Optimisation énergétique de l'étage d'adaptation électronique dédié à la conversion photovota{"\i}que. Université de Toulouse, Université Toulouse III-Paul Sabatier, 2008
62. F.A.O. Aashoor, Maximum power point tracking techniques for photovoltaic water pumping system. University of Bath, 2015
63. S. Necaibia, M.S. Kelaiaia, H. Labar, A. Necaibia, E.D. Castronuovo, Enhanced auto-scaling incremental conductance MPPT method, implemented on low-cost microcontroller and SEPIC converter. Sol. Energy **180** (2019). https://doi.org/10.1016/j.solener.2019.01.028
64. K. Ishaque, Z. Salam, A review of maximum power point tracking techniques of PV system for uniform insolation and partial shading condition. Renew. Sustain. Energy Rev. **19** (2013). https://doi.org/10.1016/j.rser.2012.11.032
65. S. Motahhir, A. El Hammoumi, A. El Ghzizal, The most used MPPT algorithms: review and the suitable low-cost embedded board for each algorithm. J. Clean. Prod. **246** (2020). https://doi.org/10.1016/j.jclepro.2019.118983
66. B. Bendib, H. Belmili, F. Krim, A survey of the most used MPPT methods: conventional and advanced algorithms applied for photovoltaic systems. Renew. Sustain. Energy Rev. **45** (2015). https://doi.org/10.1016/j.rser.2015.02.009
67. J.P. Ram, T.S. Babu, N. Rajasekar, A comprehensive review on solar PV maximum power point tracking techniques. Renew. Sustain. Energy Rev. **67** (2017). https://doi.org/10.1016/j.rser.2016.09.076
68. M. Kermadi, S. Mekhilef, Z. Salam, J. Ahmed, E.M. Berkouk, Assessment of maximum power point trackers performance using direct and indirect control methods. Int. Trans. Electr. Energy Syst. **30**(10) (2020). https://doi.org/10.1002/2050-7038.12565
69. M. Lasheen, M. Abdel-Salam, Maximum power point tracking using hill climbing and ANFIS techniques for PV applications: a review and a novel hybrid approach. Energy Convers. Manag. **171** (2018). https://doi.org/10.1016/j.enconman.2018.06.003
70. A. Loukriz, M. Haddadi, S. Messalti, Simulation and experimental design of a new advanced variable step size incremental conductance MPPT algorithm for PV systems. ISA Trans. **62** (2016). https://doi.org/10.1016/j.isatra.2015.08.006
71. S. Chtita, A. Derouich, A. El Ghzizal, S. Motahhir, An improved control strategy for charging solar batteries in off-grid photovoltaic systems. Sol. Energy **220** (2021). https://doi.org/10.1016/j.solener.2021.04.003
72. S. Motahhir, A. El Ghzizal, S. Sebti, A. Derouich, MIL and SIL and PIL tests for MPPT algorithm. Cogent Eng. **4**(1) (2017). https://doi.org/10.1080/23311916.2017.1378475
73. H.T. Yau, C.J. Lin, Q.C. Liang, PSO based PI controller design for a solar charger system. Sci. World J. **2013** (2013). https://doi.org/10.1155/2013/815280
74. X. Li, H. Wen, Y. Hu, L. Jiang, A novel beta parameter based fuzzy-logic controller for photovoltaic MPPT application. Renew. Energy **130**, 416–427 (2019). https://doi.org/10.1016/j.renene.2018.06.071
75. H.T. Yau, Q.C. Liang, C.T. Hsieh, Maximum power point tracking and optimal Li-ion battery charging control for photovoltaic charging system. Comput. Math. Appl. (2012). https://doi.org/10.1016/j.camwa.2011.12.048
76. Y. Cheddadi, F. Errahimi, N. Es-sbai, Design and verification of photovoltaic MPPT algorithm as an automotive-based embedded software. Sol. Energy **171** (2018). https://doi.org/10.1016/j.solener.2018.06.085

77. H. Fathabadi, Novel fast dynamic MPPT (maximum power point tracking) technique with the capability of very high accurate power tracking. Energy **94** (2016). https://doi.org/10.1016/j.energy.2015.10.133
78. H.D. Liu, C.H. Lin, K.J. Pai, Y.L. Lin, A novel photovoltaic system control strategies for improving hill climbing algorithm efficiencies in consideration of radian and load effect. Energy Convers. Manag. **165** (2018). https://doi.org/10.1016/j.enconman.2018.03.081
79. J. Ahmed, Z. Salam, An improved perturb and observe (P&O) maximum power point tracking (MPPT) algorithm for higher efficiency. Appl. Energy **150** (2015). https://doi.org/10.1016/j.apenergy.2015.04.006
80. H. Bounechba, A. Bouzid, H. Snani, A. Lashab, Real time simulation of MPPT algorithms for PV energy system. Int. J. Electr. Power Energy Syst. **83** (2016). https://doi.org/10.1016/j.ijepes.2016.03.041
81. A. Ostadrahimi, Y. Mahmoud, Novel spline-MPPT technique for photovoltaic systems under uniform irradiance and partial shading conditions. IEEE Trans. Sustain. Energy **12**(1) (2021). https://doi.org/10.1109/TSTE.2020.3009054
82. X. Li, H. Wen, B. Chen, S. Ding, W. Xiao, A cost-effective power ramp rate control strategy based on flexible power point tracking for photovoltaic system. Sol. Energy **208** (2020). https://doi.org/10.1016/j.solener.2020.08.044
83. Y. Zhu, M.K. Kim, H. Wen, Simulation and analysis of perturbation and observation-based self-adaptable step size maximum power point tracking strategy with low power loss for photovoltaics. Energies **12**(1) (2019). https://doi.org/10.3390/en12010092
84. N. Kumar, I. Hussain, B. Singh, B.K. Panigrahi, Framework of maximum power extraction from solar PV panel using self predictive perturb and observe algorithm. IEEE Trans. Sustain. Energy **9**(2) (2018). https://doi.org/10.1109/TSTE.2017.2764266
85. S. Mohanty, B. Subudhi, P.K. Ray, A grey wolf-assisted perturb & observe MPPT algorithm for a PV system. IEEE Trans. Energy Convers. **32**(1) (2017). https://doi.org/10.1109/TEC.2016.2633722
86. C. Manickam, G.P. Raman, G.R. Raman, S.I. Ganesan, N. Chilakapati, Fireworks enriched P&O algorithm for GMPPT and detection of partial shading in PV systems. IEEE Trans. Power Electron. **32**(6) (2017). https://doi.org/10.1109/TPEL.2016.2604279
87. K. Sundareswaran, V. Vigneshkumar, P. Sankar, S.P. Simon, P.S.R. Nayak, S. Palani, Development of an improved P&O algorithm assisted through a colony of foraging ants for MPPT in PV system. IEEE Trans. Ind. Inf. **12**(1) (2016). https://doi.org/10.1109/TII.2015.2502428
88. B. Bahmanifirouzi, E. Farjah, T. Niknam, E. Azad Farsani, Experimental verification of P&O MPPT algorithm with direct control based on fuzzy logic control using CUK converter. Iran. J. Sci. Technol. Trans. Electr. Eng. **36**(E1) (2012)
89. J. Ahmed, Z. Salam, A modified P and O maximum power point tracking method with reduced steady-state oscillation and improved tracking efficiency. IEEE Trans. Sustain. Energy **7**(4) (2016). https://doi.org/10.1109/TSTE.2016.2568043
90. J. Ahmed, Z. Salam, An enhanced adaptive P&O MPPT for fast and efficient tracking under varying environmental conditions. IEEE Trans. Sustain. Energy **9**(3) (2018). https://doi.org/10.1109/TSTE.2018.2791968
91. S.K. Kollimalla, M.K. Mishra, A novel adaptive P&O mppt algorithm considering sudden changes in the irradiance. IEEE Trans. Energy Convers. **29**(3) (2014). https://doi.org/10.1109/TEC.2014.2320930
92. M.A. Elgendy, B. Zahawi, D.J. Atkinson, Assessment of perturb and observe MPPT algorithm implementation techniques for PV pumping applications. IEEE Trans. Sustain. Energy **3**(1) (2012). https://doi.org/10.1109/TSTE.2011.2168245

References

93. O. Wasynczuk, Dynamic behavior of a class of photovoltaic power systems. IEEE Trans. Power Appar. Syst. **PAS-102**(9) (1983). https://doi.org/10.1109/TPAS.1983.318109
94. F. Liu, S. Duan, F. Liu, B. Liu, Y. Kang, A variable step size INC MPPT method for PV systems. IEEE Trans. Ind. Electron. **55**(7) (2008). https://doi.org/10.1109/TIE.2008.920550
95. T.K. Soon, S. Mekhilef, A fast-converging MPPT technique for photovoltaic system under fast-varying solar irradiation and load resistance. IEEE Trans. Ind. Inform. **11**(1) (2015). https://doi.org/10.1109/TII.2014.2378231
96. K.S. Tey, S. Mekhilef, Modified incremental conductance MPPT algorithm to mitigate inaccurate responses under fast-changing solar irradiation level. Sol. Energy **101** (2014). https://doi.org/10.1016/j.solener.2014.01.003
97. K.H. Hussein, I. Muta, T. Hoshino, M. Osakada, Maximum photovoltaic power tracking: an algorithm for rapidly changing atmospheric conditions. IEE Proc. Gener. Transm. Distrib. **142**(1) (1995). https://doi.org/10.1049/ip-gtd:19951577
98. Q. Mei, M. Shan, L. Liu, J.M. Guerrero, A novel improved variable step-size incremental-resistance MPPT method for PV systems. IEEE Trans. Ind. Electron. **58**(6) (2011). https://doi.org/10.1109/TIE.2010.2064275
99. B.N. Alajmi, K.H. Ahmed, S.J. Finney, B.W. Williams, Fuzzy-logic-control approach of a modified hill-climbing method for maximum power point in microgrid standalone photovoltaic system. IEEE Trans. Power Electron. **26**(4) (2011). https://doi.org/10.1109/TPEL.2010.2090903
100. W. Zhu, L. Shang, P. Li, H. Guo, Modified hill climbing MPPT algorithm with reduced steady-state oscillation and improved tracking efficiency. J. Eng. **17**, 2018 (2018). https://doi.org/10.1049/joe.2018.8337
101. F. Liu, Y. Kang, Z. Yu, S. Duan, Comparison of P&O and hill climbing MPPT methods for grid-connected PV converter, in *3rd IEEE Conference on Industrial Electronics and Applications, ICIEA 2008* (2008). https://doi.org/10.1109/ICIEA.2008.4582626
102. M. Veerachary, T. Senjyu, K. Uezato, Maximum power point tracking control of IDB converter supplied PV system. IEE Proc. Electr. Power Appl. (2001). https://doi.org/10.1049/ip-epa:20010656
103. A. Belkaid, I. Colak, O. Isik, Photovoltaic maximum power point tracking under fast varying of solar radiation. Appl. Energy **179** (2016). https://doi.org/10.1016/j.apenergy.2016.07.034
104. Y.P. Huang, A rapid maximum power measurement system for high-concentration photovoltaic modules using the fractional open-circuit voltage technique and controllable electronic load. IEEE J. Photovoltaics **4**(6) (2014). https://doi.org/10.1109/JPHOTOV.2014.2351613
105. Y. Zhihao, W. Xiaobo, Compensation loop design of a photovoltaic system based on constant voltage MPPT, in *Asia-Pacific Power and Energy Engineering Conference* (APPEEC, 2009). https://doi.org/10.1109/APPEEC.2009.4918231
106. M. Hlaili, H. Mechergui, Comparison of different MPPT algorithms with a proposed one using a power estimator for grid connected PV systems. Int. J. Photoenergy **2016** (2016). https://doi.org/10.1155/2016/1728398
107. Z. Salam, J. Ahmed, B.S. Merugu, The application of soft computing methods for MPPT of PV system: a technological and status review. Appl. Energy **107**, 135–148 (Elsevier Ltd, 2013). https://doi.org/10.1016/j.apenergy.2013.02.008
108. M. Farhat, O. Barambones, L. Sbita, Efficiency optimization of a DSP-based standalone PV system using a stable single input fuzzy logic controller. Renew. Sustain. Energy Rev. **49** (2015). https://doi.org/10.1016/j.rser.2015.04.123
109. U. Yilmaz, A. Kircay, S. Borekci, PV system fuzzy logic MPPT method and PI control as a charge controller. Renew. Sustain. Energy Rev. **81** (2018). https://doi.org/10.1016/j.rser.2017.08.048

110. S. Ouchen, A. Betka, S. Abdeddaim, A. Menadi, Fuzzy-predictive direct power control implementation of a grid connected photovoltaic system, associated with an active power filter. Energy Convers. Manag. **122** (2016). https://doi.org/10.1016/j.enconman.2016.06.018
111. M. Nabipour, M. Razaz, S.G. Seifossadat, S.S. Mortazavi, A new MPPT scheme based on a novel fuzzy approach. Renew. Sustain. Energy Rev. **74** (2017). https://doi.org/10.1016/j.rser.2017.02.054
112. K. Loukil, H. Abbes, H. Abid, M. Abid, A. Toumi, Design and implementation of reconfigurable MPPT fuzzy controller for photovoltaic systems. Ain Shams Eng. J. **11**(2) (2020). https://doi.org/10.1016/j.asej.2019.10.002
113. T.H. Kwan, X. Wu, Maximum power point tracking using a variable antecedent fuzzy logic controller. Sol. Energy **137** (2016). https://doi.org/10.1016/j.solener.2016.08.008
114. S. Farajdadian, S.M.H. Hosseini, Design of an optimal fuzzy controller to obtain maximum power in solar power generation system. Sol. Energy **182** (2019). https://doi.org/10.1016/j.solener.2019.02.051
115. Syafaruddin, E. Karatepe, T. Hiyama, Artificial neural network-polar coordinated fuzzy controller based maximum power point tracking control under partially shaded conditions. IET Renew. Power Gener. **3**(2) (2009). https://doi.org/10.1049/iet-rpg:20080065
116. A. Messai, A. Mellit, A. Guessoum, S.A. Kalogirou, Maximum power point tracking using a GA optimized fuzzy logic controller and its FPGA implementation. Sol. Energy **85**(2) (2011). https://doi.org/10.1016/j.solener.2010.12.004
117. M.A. Aredes, B.W. França, M. Aredes, Fuzzy adaptive P & O control for MPPT of a photovoltaic module. J. Power Energy Eng. **2**(4) (2014). https://doi.org/10.4236/jpee.2014.24018
118. M.N. Ali, K. Mahmoud, M. Lehtonen, M.M.F. Darwish, An efficient fuzzy-logic based variable-step incremental conductance MPPT method for grid-connected PV systems. IEEE Access **9**(2021). https://doi.org/10.1109/ACCESS.2021.3058052.
119. T. Zhu, J. Dong, X. Li, S. Ding, A comprehensive study on maximum power point tracking techniques based on fuzzy logic control for solar photovoltaic systems. Front. Energy Res. **9** (2021). https://doi.org/10.3389/fenrg.2021.727949
120. H. Rezk, E.S. Hasaneen, A new MATLAB/simulink model of triple-junction solar cell and MPPT based on artificial neural networks for photovoltaic energy systems. Ain Shams Eng. J. **6**(3) (2015). https://doi.org/10.1016/j.asej.2015.03.001
121. T. Hiyama, S. Kouzuma, T. Imakubo, Identification of optimal operating point of PV modules using neural network for real time maximum power tracking control. IEEE Trans. Energy Convers. **10**(2) (1995). https://doi.org/10.1109/60.391904
122. H. Boumaaraf, A. Talha, O. Bouhali, A three-phase NPC grid-connected inverter for photovoltaic applications using neural network MPPT. Renew. Sustain. Energy Rev. **49** (2015). https://doi.org/10.1016/j.rser.2015.04.066
123. M. Adly, H. El-Sherif, M. Ibrahim, Maximum power point tracker for a PV cell using a fuzzy agent adapted by the fractional open circuit voltage technique, in *IEEE International Conference on Fuzzy Systems* (2011). https://doi.org/10.1109/FUZZY.2011.6007697
124. A. Reza Reisi, M. Hassan Moradi, S. Jamasb, Classification and comparison of maximum power point tracking techniques for photovoltaic system: a review. Renew. Substain. Energy Rev. **19** (2013). https://doi.org/10.1016/j.rser.2012.11.052
125. K. Ishaque, Z. Salam, A. Shamsudin, M. Amjad, A direct control based maximum power point tracking method for photovoltaic system under partial shading conditions using particle swarm optimization algorithm. Appl. Energy **99** (2012). https://doi.org/10.1016/j.apenergy.2012.05.026

126. M. Kermadi, E.M. Berkouk, Artificial intelligence-based maximum power point tracking controllers for photovoltaic systems: comparative study, in *Renewable and Sustainable Energy Reviews*, vol. 69 (Elsevier Ltd, 2017), pp. 369–386. https://doi.org/10.1016/j.rser.2016.11.125
127. M. Miyatake, M. Veerachary, F. Toriumi, N. Fujii, H. Ko, Maximum power point tracking of multiple photovoltaic arrays: a PSO approach. IEEE Trans. Aerosp. Electron. Syst. **47**(1) (2011). https://doi.org/10.1109/TAES.2011.5705681
128. K. Ishaque, Z. Salam, M. Amjad, S. Mekhilef, An improved particle swarm optimization (PSO)-based MPPT for PV with reduced steady-state oscillation. IEEE Trans. Power Electron. **27**(8) (2012). https://doi.org/10.1109/TPEL.2012.2185713
129. J.J. Soon, K.S. Low, Photovoltaic model identification using particle swarm optimization with inverse barrier constraint. IEEE Trans. Power Electron. **27**(9) (2012) https://doi.org/10.1109/TPEL.2012.2188818
130. Y.H. Liu, S.C. Huang, J.W. Huang, W.C. Liang, A particle swarm optimization-based maximum power point tracking algorithm for PV systems operating under partially shaded conditions. IEEE Trans. Energy Convers. **27**(4) (2012). https://doi.org/10.1109/TEC.2012.2219533
131. S.M. Mirhassani, S.Z.M. Golroodbari, S.M.M. Golroodbari, S. Mekhilef, An improved particle swarm optimization based maximum power point tracking strategy with variable sampling time. Int. J. Electr. Power Energy Syst. **64** (2015). https://doi.org/10.1016/j.ijepes.2014.07.074
132. F. Belhachat, C. Larbes, A review of global maximum power point tracking techniques of photovoltaic system under partial shading conditions. Renew. Sustain. Energy Rev. **92**, 513–553 (2018). https://doi.org/10.1016/J.RSER.2018.04.094
133. A.A.S. Mohamed, A. Berzoy, O.A. Mohammed, Design and hardware implementation of FL-MPPT control of PV systems based on GA and small-signal analysis. IEEE Trans. Sustain. Energy **8**(1) (2017). https://doi.org/10.1109/TSTE.2016.2598240
134. J. Ahmed, Z. Salam, A maximum power point tracking (MPPT) for PV system using cuckoo search with partial shading capability. Appl. Energy **119** (2014). https://doi.org/10.1016/j.apenergy.2013.12.062
135. C. Gonzalez-Castano, C. Restrepo, S. Kouro, J. Rodriguez, MPPT algorithm based on artificial bee colony for PV system. IEEE Access **9** (2021). https://doi.org/10.1109/ACCESS.2021.3066281
136. M. Rajasekaran, A.C. Vaithlingam, Maximum power point tracking for PV array based on ant colony optimization under uniform and non-uniform irradiance. Int. J. Intellect. Adv. Res. Eng. Comput. **5**(2) (2017)
137. S. Mohanty, B. Subudhi, P.K. Ray, A new MPPT design using grey Wolf optimization technique for photovoltaic system under partial shading conditions. IEEE Trans. Sustain. Energy **7**(1) (2016). https://doi.org/10.1109/TSTE.2015.2482120
138. D.F. Teshome, C.H. Lee, Y.W. Lin, K.L. Lian, A modified firefly algorithm for photovoltaic maximum power point tracking control under partial shading. IEEE J. Emerg. Sel. Top. Power Electron. **5**(2) (2017). https://doi.org/10.1109/JESTPE.2016.2581858
139. J. Liu, S. Zheng, Y. Tan, The improvement on controlling exploration and exploitation of firework algorithm. Lecture Notes in Computer Science (including subseries Lecture Notes in Artificial Intelligence and Lecture Notes in Bioinformatics) (2013). https://doi.org/10.1007/978-3-642-38703-6_2
140. K.H. Chao, M.N. Rizal, A hybrid mppt controller based on the genetic algorithm and ant colony optimization for photovoltaic systems under partially shaded conditions. Energies **14**(10) (2021). https://doi.org/10.3390/en14102902
141. K. Sundareswaran, V. Vignesh kumar, S. Palani, Application of a combined particle swarm optimization and perturb and observe method for MPPT in PV systems under partial shading conditions. Renew. Energy **75** (2015). https://doi.org/10.1016/j.renene.2014.09.044

142. K.L. Lian, J.H. Jhang, I.S. Tian, A maximum power point tracking method based on perturb-and-observe combined with particle swarm optimization. IEEE J. Photovoltaics **4**(2) (2014). https://doi.org/10.1109/JPHOTOV.2013.2297513
143. Ö. Çelik, A. Teke, A Hybrid MPPT method for grid connected photovoltaic systems under rapidly changing atmospheric conditions. Electr. Power Syst. Res. **152** (2017). https://doi.org/10.1016/j.epsr.2017.07.011
144. H.M. El-Helw, A. Magdy, M.I. Marei, A hybrid maximum power point tracking technique for partially shaded photovoltaic arrays. IEEE Access **5** (2017). https://doi.org/10.1109/ACCESS.2017.2717540
145. M.A. Danandeh, S.M.G. Mousavi, A new architecture of INC-fuzzy hybrid method for tracking maximum power point in PV cells. Sol. Energy **171** (2018). https://doi.org/10.1016/j.solener.2018.06.098
146. T. Radjai, L. Rahmani, S. Mekhilef, J.P. Gaubert, Implementation of a modified incremental conductance MPPT algorithm with direct control based on a fuzzy duty cycle change estimator using dSPACE. Sol. Energy **110** (2014). https://doi.org/10.1016/j.solener.2014.09.014
147. J.Y. Shi et al. Combining incremental conductance and firefly algorithm for tracking the global MPP of PV arrays. J. Renew. Sustain. Energy **9**(2) (2017). https://doi.org/10.1063/1.4977213
148. S.A. Rizzo, G. Scelba, A hybrid global MPPT searching method for fast variable shading conditions. J. Clean. Prod. **298**, 126775 (2021). https://doi.org/10.1016/J.JCLEPRO.2021.126775
149. S.A. Rizzo, N. Salerno, G. Scelba, A. Sciacca, Enhanced hybrid global MPPT algorithm for PV systems operating under fast-changing partial shading conditions. Int. J. Renew. Energy Res. **8**(1) (2018)
150. M.N. Ali, K. Mahmoud, M. Lehtonen, M.M.F. Darwish, Promising mppt methods combining metaheuristic, fuzzy-logic and ANN techniques for grid-connected photovoltaic. Sensors (Switzerland) **21**(4) (2021). https://doi.org/10.3390/s21041244
151. R.K. Kharb, S.L. Shimi, S. Chatterji, M.F. Ansari, Modeling of solar PV module and maximum power point tracking using ANFIS. Renew. Sustain. Energy Rev. **33** (2014). https://doi.org/10.1016/j.rser.2014.02.014
152. B. Laxman, A. Annamraju, N.V. Srikanth, A grey wolf optimized fuzzy logic based MPPT for shaded solar photovoltaic systems in microgrids. Int. J. Hydrogen Energy **46**(18), 10653–10665 (2021). https://doi.org/10.1016/J.IJHYDENE.2020.12.158
153. B. Babes, A. Boutaghane, N. Hamouda, A novel nature-inspired maximum power point tracking (MPPT) controller based on ACO-ANN algorithm for photovoltaic (PV) system fed arc welding machines. Neural Comput. Appl. (2021). https://doi.org/10.1007/s00521-021-06393-w
154. L.G.K. Chai, L. Gopal, F.H. Juwono, C.W.R. Chiong, H.C. Ling, T.A. Basuki, A novel global MPPT technique using improved PS-FW algorithm for PV system under partial shading conditions. Energy Convers. Manag. **246**, 114639 (2021). https://doi.org/10.1016/J.ENCONMAN.2021.114639
155. F. Zhao, Z. Yao, J. Luan, X. Song, A novel fused optimization algorithm of genetic algorithm and ant colony optimization. Math. Probl. Eng. **2016** (2016). https://doi.org/10.1155/2016/2167413
156. M.G. Batarseh, M.E. Za'ter, Hybrid maximum power point tracking techniques: a comparative survey, suggested classification and uninvestigated combinations. Solar Energy **169** (2018). https://doi.org/10.1016/j.solener.2018.04.045
157. D. Akinyele, J. Belikov, Y. Levron, Battery storage technologies for electrical applications: impact in stand-alone photovoltaic systems. Energies **10**(11) (2017). https://doi.org/10.3390/en10111760

158. H. Chen, T.N. Cong, W. Yang, C. Tan, Y. Li, Y. Ding, Progress in electrical energy storage system: a critical review. Progress Nat. Sci. **19**(3) (2009). https://doi.org/10.1016/j.pnsc.2008.07.014
159. B. Zakeri, S. Syri, Electrical energy storage systems: a comparative life cycle cost analysis. Renew. Sustain. Energy Rev. **42** (2015). https://doi.org/10.1016/j.rser.2014.10.011
160. A. Jossen, J. Garche, D.U. Sauer, Operation conditions of batteries in PV applications. Sol. Energy **76**(6), 759–769 (2004). https://doi.org/10.1016/J.SOLENER.2003.12.013
161. K.C. Divya, J. Østergaard, Battery energy storage technology for power systems-an overview. Electr. Power Syst. Res. **79**(4) (2009). https://doi.org/10.1016/j.epsr.2008.09.017
162. X. Luo, J. Wang, M. Dooner, J. Clarke, Overview of current development in electrical energy storage technologies and the application potential in power system operation. Appl. Energy **137** (2015). https://doi.org/10.1016/j.apenergy.2014.09.081
163. G. Merei, C. Berger, D.U. Sauer, Optimization of an off-grid hybrid PV–wind–diesel system with different battery technologies using genetic algorithm. Sol. Energy **97**, 460–473 (2013). https://doi.org/10.1016/J.SOLENER.2013.08.016
164. A.A. Hussein, A.A. Fardoun, Design considerations and performance evaluation of outdoor PV battery chargers. Renew. Energy **82** (2015). https://doi.org/10.1016/j.renene.2014.08.063
165. D. Spiers, Batteries in PV systems, in *Practical Handbook Photovoltaics* (2012), pp. 721–776. https://doi.org/10.1016/B978-0-12-385934-1.00022-2
166. Z.M. Salameh, M.A. Casacca, W.A. Lynch, A mathematical model for lead-acid batteries. IEEE Trans. Energy Convers. **7**(1) (1992). https://doi.org/10.1109/60.124547
167. S.O. Amrouche, D. Rekioua, T. Rekioua, S. Bacha, Overview of energy storage in renewable energy systems. Int. J. Hydrogen Energy **41**(45), 20914–20927 (2016). https://doi.org/10.1016/J.IJHYDENE.2016.06.243
168. B. Bogno et al., Improvement of safety, longevity and performance of lead acid battery in off-grid PV systems (2016). https://doi.org/10.1016/j.ijhydene.2016.12.011
169. N.K.C. Nair, N. Garimella, Battery energy storage systems: assessment for small-scale renewable energy integration. Energy Build. **42**(11), 2124–2130 (2010). https://doi.org/10.1016/J.ENBUILD.2010.07.002
170. X. Ge, F.W. Ahmed, A. Rezvani, N. Aljojo, S. Samad, L.K. Foong, Implementation of a novel hybrid BAT-Fuzzy controller based MPPT for grid-connected PV-battery system. Control Eng. Pract. **98** (2020). https://doi.org/10.1016/j.conengprac.2020.104380
171. R.M. Herritty, J. Midolo, Nickel cadmium batteries for photovoltaic applications, in *Proceedings of the Annual Battery Conference on Applications and Advances* (1998). https://doi.org/10.1109/bcaa.1998.653876
172. M.A. Hannan, M.M. Hoque, A. Mohamed, A. Ayob, Review of energy storage systems for electric vehicle applications: issues and challenges. Renew. Sustain. Energy Rev. **69** (2017). https://doi.org/10.1016/j.rser.2016.11.171
173. A.H. Fathima, K. Palanisamy, Battery energy storage applications in wind integrated systems—a review, in *2014 International Conference on Smart Electric Grid, ISEG 2014* (2015). https://doi.org/10.1109/ISEG.2014.7005604
174. IRENA, Battery storage for renewables: market status and technology outlook. Irena (2015)
175. D. Magnor, J.B. Gerschler, M. Ecker, P. Merk, D.U. Sauer, Concept of a battery aging model for lithium-ion batteries considering the lifetime dependency on the operation strategy, in *24th European Photovoltaic Solar Energy Conference*, 21–25 Sept 2009

Modeling and Simulation of Standalone Solar Photovoltaic Systems

3.1 Introduction

The photovoltaic effect, first discovered by the French physicist Alexandre Edmond Becquerel in 1839, offers an exceptional opportunity to harness clean, renewable, and sustainable energy from sunlight. This marked the initial transformation of solar energy into electrical energy.

The most commonly used semiconductor material for manufacturing solar cells worldwide is silicon (SI), as illustrated in Fig. 3.1a. The Si atom, with its four valence electrons, can form bonds with other atoms. In its crystalline structure, each Si atom shares its valence electrons with four neighboring Si atoms, creating a stable lattice, as shown in Fig. 3.1b.

Fundamentally, a basic solar cell operates as a simple diode, as depicted in Fig. 3.2. It consists of two thin layers of Si semiconductor material: the upper layer is N-type doped (typically with phosphorus), and the lower layer is P-type doped (typically with boron). This configuration creates a PN junction between them. When photons from sunlight strike the front surface of the semiconductor, they excite the valence electrons, resulting in the generation of electron-hole pairs. This leads to the creation of a potential difference between the two layers, generating an electrical current in the solar cell [1].

As illustrated in Fig. 3.3, three types of silicon-based PV cells are most commonly used, compared to other PV technologies [2, 3]:

- Poly-crystalline
- Mono-crystalline or Single-crystal
- Thin-film amorphous.

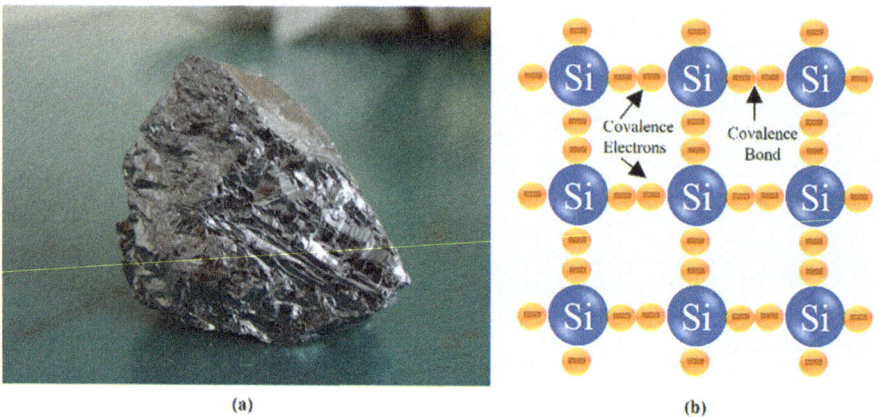

Fig. 3.1 Silicon (Si); **a** material and **b** crystalline structure of atoms

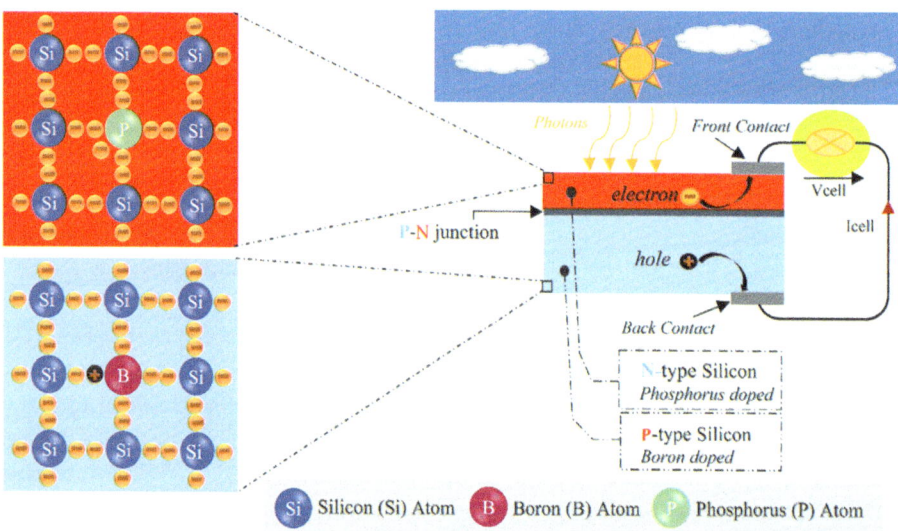

Fig. 3.2 A graphical representation of the operation of a PV cell

In addition to silicon-based PV technologies, there are other types such as cadmium telluride (CdTe), gallium arsenide (GaAs), copper indium selenide (CIS), and copper indium gallium selenide (CIGS). These photovoltaic technologies are generally classified into three generations [2]:

1. **First Generation**: Silicon wafer-based PV cells (Poly-crystalline and Mono-crystalline).

3.1 Introduction

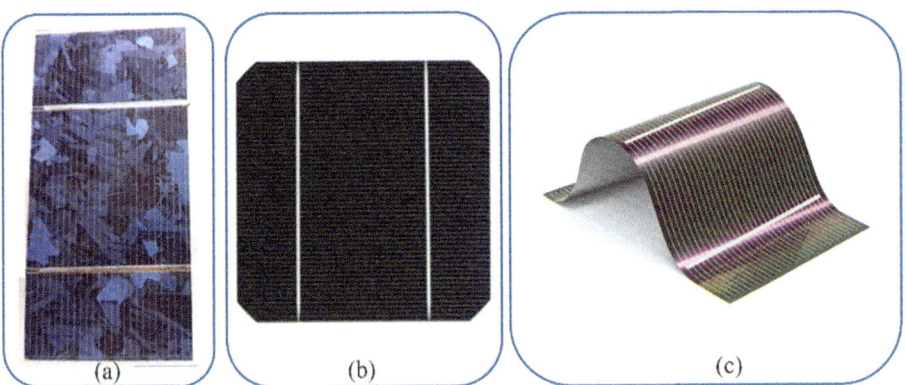

Fig. 3.3 Three types of PV cells based on Si material; **a** poly-crystalline, **b** mono-crystalline, and **c** thin-film amorphous

2. **Second Generation**: Thin-film solar cells (including amorphous silicon and CdTe).
3. **Third Generation**: Emerging technologies such as organic photovoltaics and multi-junction cells.

Several parameters influence the current versus voltage (I–V) and power versus voltage (P–V) characteristics of PV cells, modules, and arrays. These parameters can be categorized into:

- **Meteorological Factors**: Such as solar irradiance, ambient temperature, and partial shading.
- **Electrical Factors**: Such as ideality factor, saturation current, and shunt and series resistances.

To demonstrate the impact of these parameters and validate their effects, this chapter presents various P–V and I–V characteristic plots under different conditions. Additionally, this chapter focuses on the modeling and simulation of a standalone PV system. In the modeling section, the two primary components of the standalone PV system—namely, the solar PV module and the MPPT system—will be modeled using robust tools like MATLAB/Simulink and Proteus software.

The simulation section will involve the implementation and simulation of the complete standalone PV system using conventional MPPT algorithms such as P&O and INC methods. This will provide a comprehensive understanding of the behavior and performance of the standalone PV system under various operating conditions.

3.2 Modeling of a Standalone Solar Photovoltaic System

As discussed in the previous chapter, solar PV systems can be installed in two primary configurations: off-grid systems and grid-connected (on-grid) systems. In this thesis, the focus is on the standalone PV system installation.

Standalone solar PV systems are utilized in various applications, including street lighting, solar-powered vehicles, power supply for remote areas, and space satellite systems [4]. These systems can be categorized into three main types: unregulated systems, and regulated systems with or without battery storage [5].

Unregulated standalone PV systems are the simplest configuration, where the PV module or array is directly connected to the available DC loads. In contrast, regulated standalone PV systems include a power conversion stage, such as a DC-DC converter and a MPPT controller, to manage the power flow between the PV module and the loads [6]. Figure 3.4 illustrates the different installation types of standalone PV systems designed to supply power to DC loads.

3.2.1 Modeling of a Solar Photovoltaic Module

3.2.1.1 Photovoltaic Cell

A photovoltaic cell is the fundamental unit responsible for the direct conversion of sunlight into electrical energy through the photovoltaic effect. It serves as the basic building block for constructing PV modules or arrays. As mentioned in the introduction, a PV cell consists of two thin layers of semiconductor material, typically silicon, one doped as

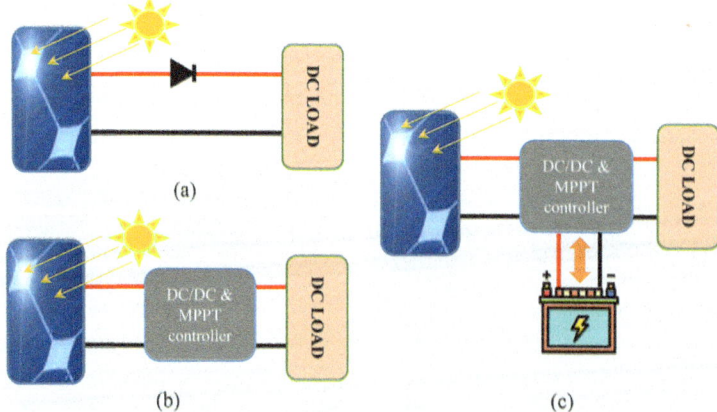

Fig. 3.4 Different types of standalone PV system installations: **a** unregulated, **b** regulated without batteries, and **c** regulated with batteries

3.2 Modeling of a Standalone Solar Photovoltaic System

N-type and the other as P-type. Essentially, a PV cell functions similarly to a diode, as commonly used in electronics.

3.2.1.2 Photovoltaic Module

Since a single PV cell produces a relatively low voltage (approximately 0.5 V), multiple cells are connected in various configurations (series and/or parallel) to form a PV module capable of delivering the desired power output. Series connections increase the overall voltage, while parallel connections increase the current. However, shading of individual cells within a module can lead to a significant increase in voltage across the shaded cell, potentially causing a "hot spot" effect that could damage the cell. To mitigate this issue, bypass diodes are employed to protect the solar cells, as illustrated in Fig. 3.5.

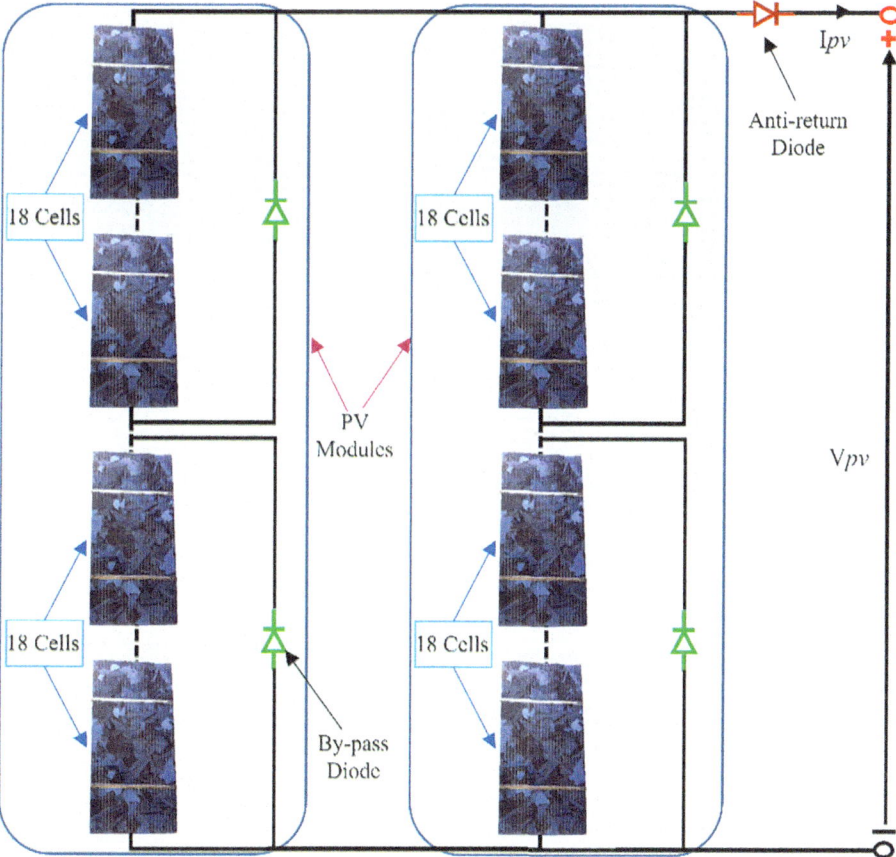

Fig. 3.5 By-pass and anti-return diodes for PV module protection

Fig. 3.6 Configuration types and relationships between PV cell, module and array

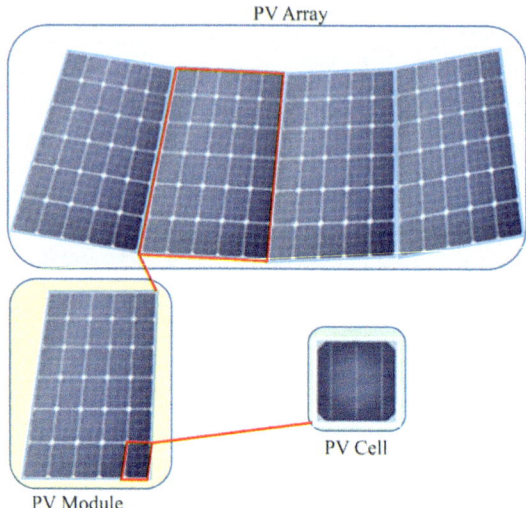

Additionally, when PV modules are connected in parallel or interfaced with batteries, another potential problem arises that could damage the modules. To prevent this, anti-return diodes are used, as depicted in Fig. 3.5, to protect the system from reverse currents.

3.2.1.3 Photovoltaic Array

When higher power demands are required, a single PV module cannot produce sufficient energy to meet the load requirements. In such cases, multiple PV modules are connected in series and/or parallel to form a photovoltaic (PV) array. Figure 3.6 illustrates the configuration types and the relationships between a PV cell, module, and array.

3.2.1.4 Equivalent Circuit and Mathematical Model

Modeling a PV cell is essential to predict its behavior under varying climatic conditions and to generate its characteristic current-voltage (I–V) and power-voltage (P–V) curves [7]. This modeling process is also critical for evaluating the influence of different parameters on the operation of PV cells [8].

Typically, a PV cell is represented by an equivalent electrical circuit consisting of a light-generated current source, one, two, or three anti-parallel diodes (D), an internal series resistance (R_s), and a shunt/parallel resistance (R_p) [9]. Figure 3.7 presents the common electrical circuit models for a solar cell that have been proposed in the literature, including the ideal, single-diode, two-diode, and three-diode models [10–12]. However, the single-diode model, also known as the five-parameter (5-p) model (depicted in Fig. 3.7b), is the most widely used due to its simplicity and the high level of accuracy it provides in simulation results compared to the more complex two- and three-diode models [13].

3.2 Modeling of a Standalone Solar Photovoltaic System

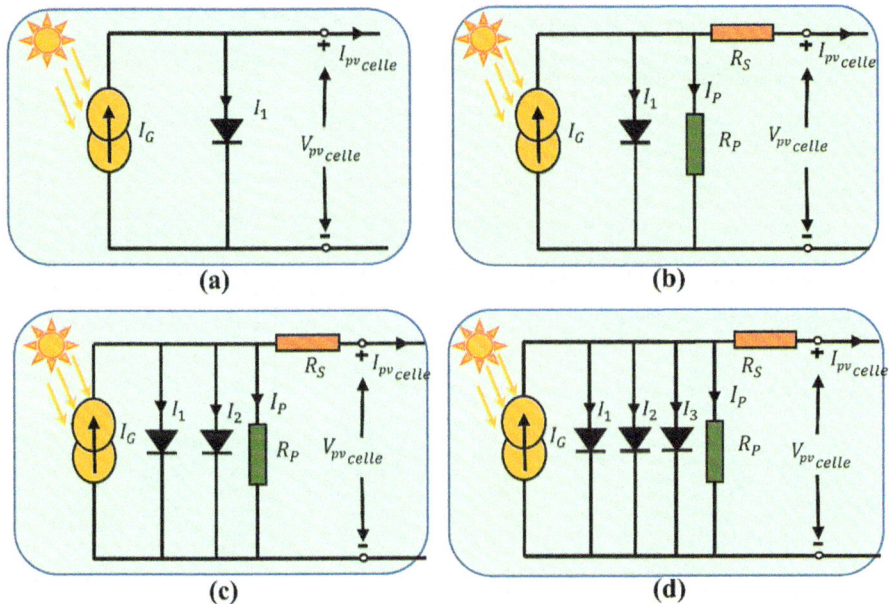

Fig. 3.7 Equivalent solar cell electrical circuit; **a** ideal model, **b** one-diode model, **c** two-diode model, and **d** three-diode model

The mathematical expressions describing the relationship between the solar cell terminal current and voltage for the equivalent circuit models—such as the ideal, single-diode, two-diode, and three-diode models—are given by (3.1), (3.2), (3.3), and (3.4), respectively, as [14, 15]:

$$I_{pv(cell)} = I_G - I_1 = I_G - I_{s1}[\exp\left(\frac{V_{pv}}{A_1 \times V_{t1}}\right) - 1] \quad (3.1)$$

$$I_{pv(cell)} = I_G - I_1 - I_p = I_G - I_{s1}\left[\exp\left(\frac{V_{pv} + I_{pv} \times R_S}{A_1 \times V_{t1}}\right) - 1\right] - \frac{V_{pv} + I_{pv} \times R_S}{R_P} \quad (3.2)$$

$$\begin{aligned} I_{pv(cell)} &= I_G - I_1 - I_2 - I_p \\ &= I_G - I_{s1}\left[\exp\left(\frac{V_{pv} + I_{pv} \times R_S}{A_1 \times V_{t1}}\right) - 1\right] - I_{s2}\left[\exp\left(\frac{V_{pv} + I_{pv} \times R_S}{A_2 \times V_{t2}}\right) - 1\right] \\ &\quad - \frac{V_{pv} + I_{pv} \times R_S}{R_P} \end{aligned} \quad (3.3)$$

$$I_{pv(cell)} = I_G - I_1 - I_2 - I_3 - I_p$$

$$= I_G - I_{s1}\left[\exp\left(\frac{V_{pv} + I_{pv} \times R_S}{A_1 \times V_{t1}}\right) - 1\right] - I_{s2}\left[\exp\left(\frac{V_{pv} + I_{pv} \times R_S}{A_2 \times V_{t2}}\right) - 1\right]$$
$$- I_{s3}\left[\exp\left(\frac{V_{pv} + I_{pv} \times R_S}{A_3 \times V_{t3}}\right) - 1\right] - \frac{V_{pv} + I_{pv} \times R_S}{R_P} \quad (3.4)$$

where:

$$I_G = \left(\frac{G}{1000}\right)[I_{sj} + k_i(T - T_r)] \quad (3.5)$$

$$I_{sj}(T) = I_{soj} \times \left(\frac{T}{T_r}\right)^3 \times \exp\left[\left(-\frac{q \times E_g}{A_j \times K}\right) \times \left(\frac{1}{T_r} - \frac{1}{T}\right)\right] \quad (3.6)$$

$$I_{soj} = \frac{I_{sc}}{\exp\left(\frac{qV_{oc}}{A_j \times V_{tj}}\right) - 1} \quad (3.7)$$

$$V_{tj} = \frac{K \times T}{q} \quad (3.8)$$

where:

- j denotes the jth diode in the equivalent circuit model of the solar cell.
- I_1, I_2, and I_3 represent the currents flowing through the corresponding diodes.
- I_{s1}, I_{s2}, and I_{s3} are the reverse saturation currents at temperature T [A].
- K is Boltzmann's constant (8.617×10^{-5} [eV/K]).
- q is the elementary charge (1.6×10^{-19} [C]).
- k_i is the short-circuit current temperature coefficient.
- E_g represents the bandgap energy of the semiconductor material.
- V_{t1}, V_{t2}, and V_{t3} are the thermal voltages of the solar cell.
- T and T_r refer to the actual and reference temperatures of the solar cell, respectively.
- A_1, A_2, and A_3 are dimensionless ideality factors of the junction material.

Consequently, the electrical equivalent circuit of a typical photovoltaic module can be obtained by arranging N_S series and N_P parallel branches, as shown in Fig. 3.8. The mathematical expression for its output current can be given by Eq. (3.9) as follows [10, 17]:

$$I_{pv(module)} = N_p \times I_G - N_p \times I_{s1}\left[xp\left(\frac{V_{pv} + I_{pv} \times R_S}{N_s \times V_{t1}}\right) - 1\right] - N_p \times \frac{V_{pv} + I_{pv} \times R_S}{N_s \times R_P} \quad (3.9)$$

It should be noted that both the graphical and mathematical model of the PV module, as illustrated in Fig. 3.8 and described by Eq. (3.9), are based on the single-diode model.

3.2 Modeling of a Standalone Solar Photovoltaic System

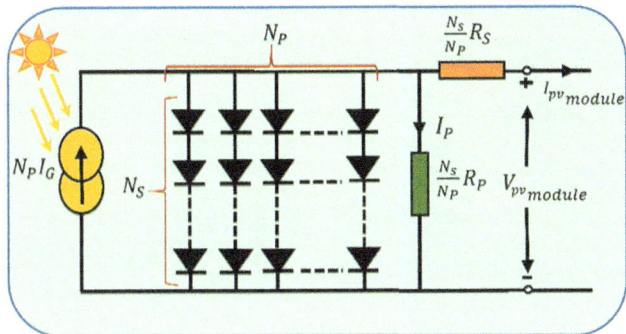

Fig. 3.8 Equivalent PV module electrical circuit with NS Series and NP parallel branches

Table 3.1 Parameters of equivalent electrical circuit models of a PV cell and their meaning

Models	Number of parameters	Parameters	The meaning of parameters
Ideal model	3	I_G, I_{s1}, A_1	I_G is photo generated current
Single-diode model	5	I_G, I_1, A_1, R_s, R_p	I_{s1}, I_{s2} and I_{s3} are saturation current at T_r of the first, second and third diode, respectively. A_1, A_2, and A_3 are the ideality factor of the first, second and third diode, respectively. R_s is the series resistance. R_p is the shunt resistance
Two-diode model	7	$I_G, I_{s1}, A_1, A_2, R_s, R_p$	
Three-diode model	9	$I_G, I_{s1}, I_{s2}, I_{s3}, A_1, A_2, A_3, R_s, R_p$	

Table 3.1 provides the definitions and meanings of the parameters used in the equivalent electrical circuit models of a PV cell.

3.2.1.5 Parameters of The Solar PV Module

A typical solar PV module can be characterized by five fundamental parameters: short-circuit current, open-circuit voltage, MPP, fill factor, and efficiency [18, 19].

Short-Circuit Current

As illustrated in Fig. 3.9, the short-circuit current (I_{SC}) is the point where the current-voltage (I–V) curve intersects the vertical axis. It represents the maximum current that a PV module can generate when its terminals are short-circuited ($V_{PV} = 0$).

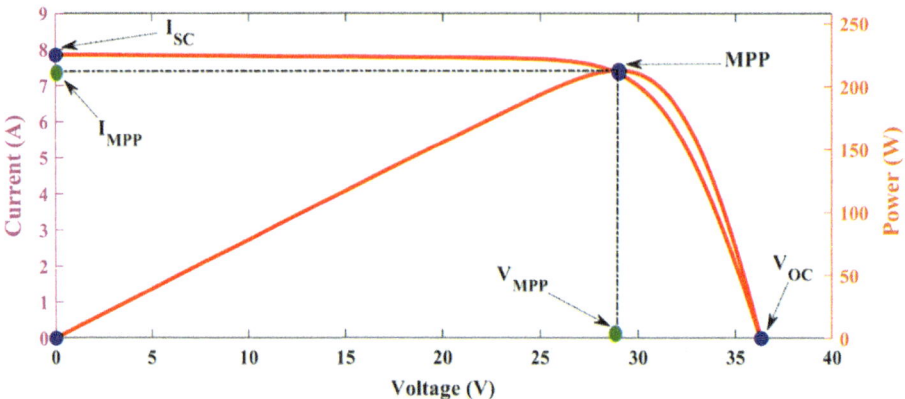

Fig. 3.9 A graphical illustration of the main parameters of a PV module

Open-Circuit Voltage

The open-circuit voltage (V_{OC}) is the maximum voltage observed at the terminals of the PV module when no current flows (under open-circuit conditions). It is represented in Fig. 3.9 as the point where the I–V curve intersects the horizontal axis.

Maximum Power Point (MPP)

The MPP is the operating condition at which the PV module delivers the highest possible power output. As shown in Fig. 3.9, it corresponds to two specific values: V_{MPP} (the optimum voltage) and I_{MPP} (the optimum current) at the MPP.

Fill Factor

The fill factor (FF) is a measure of the quality of a PV module, indicating how well the voltage and current at the maximum power point compare to the open-circuit voltage and short-circuit current. It is calculated as the ratio of the maximum power to the product of V_{OC} and I_{SC}, as shown in Eq. (3.10) [16, 18]. Typically, the FF value is less than 1; for example, for a crystalline PV module, the FF is between 0.7 and 0.9.

$$FF = \frac{P_{MPP}}{V_{OC} \times I_{SC}} = \frac{V_{MPP} \times I_{MPP}}{V_{OC} \times I_{SC}} \tag{3.10}$$

Efficiency

The efficiency (η) of a PV module is defined as the ratio of the maximum output power (P_{MPP}) to the power of the incident solar radiation (P_{in}). The efficiency is expressed by Eq. (3.11):

$$\eta = \frac{P_{MPP}}{P_{in}} = \frac{V_{MPP} \times I_{MPP}}{A \times G} \tag{3.11}$$

3.2 Modeling of a Standalone Solar Photovoltaic System

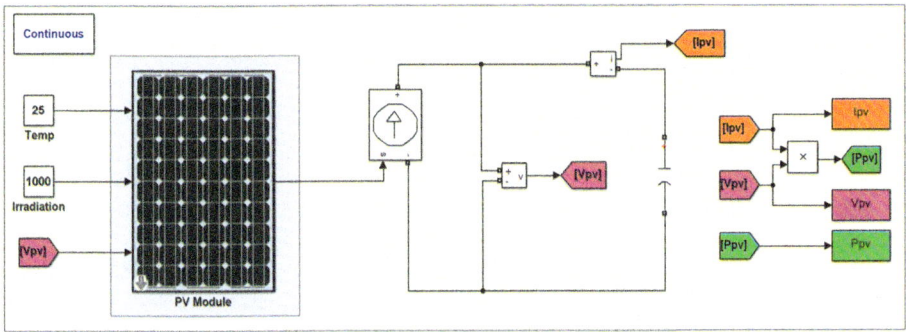

Fig. 3.10 Block diagram of the PV module developed in the Simulink environment

Here, A is the area of the PV module, and G is the irradiance. It is important to note that efficiency is typically defined under standard test conditions (STC), which assume an irradiance of 1 kW/m^2, a temperature of 25 °C, and an air mass of 1.5 [18].

3.2.1.6 Modeling of The Solar PV Module Using MATLAB/Simulink and Proteus Platforms

In this section, the modeling of a solar PV panel is conducted using two widely recognized software environments: MATLAB/Simulink and Proteus.

The modeling of the PV module in MATLAB/Simulink is based on mathematical expressions derived from the single-diode model, as discussed earlier. Three different approaches were utilized for modeling the PV module: Simulink subsystem blocks, the PV array block developed by MathWorks, and MATLAB scripting.

Simulink-Based Modeling

Figures 3.10 and 3.11 show the block diagram of the developed PV module and its contents in the Simulink platform. Figures 3.12, 3.13, 3.14, and 3.15 illustrate the contents of the subsystems shown in Fig. 3.11, where Eqs. (3.5)–(3.9), based on the single-diode model, were implemented.

PV Array Block-Based Modeling

In 2015, MathWorks, Inc. introduced a new block named "PV Array" in version R2015a of the MATLAB/Simulink software, as revealed in Fig. 3.16. As shown in Fig. 3.17 (highlighted by the red rectangle), the PV Array block includes a library reference for more than 10,000 commercially available solar modules. Furthermore, the block, depicted in Fig. 3.16, enables users to model pre-configured PV modules from the National Renewable Energy Laboratory (NREL) database, as well as to define custom PV modules [20].

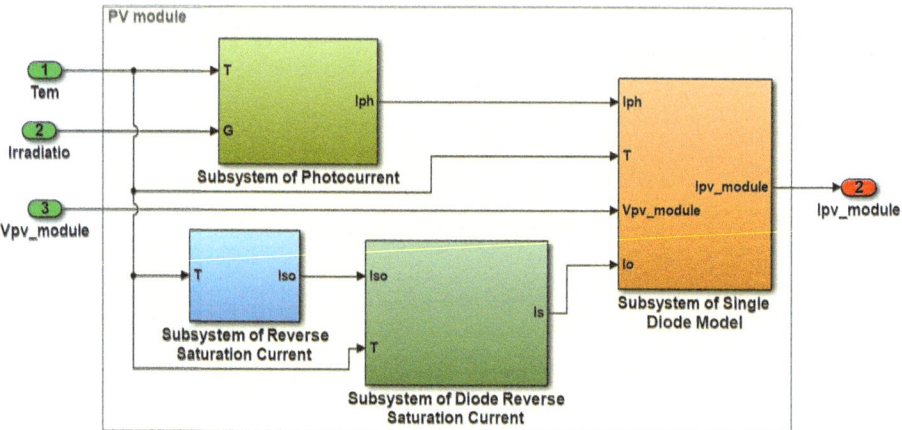

Fig. 3.11 Contents of the PV module block

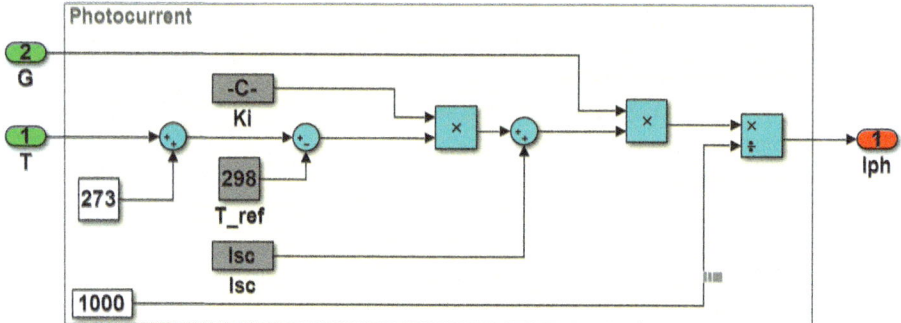

Fig. 3.12 Implementation of the photocurrent using Eq. (3.5). This corresponds to the subsystem labeled "subsystem of photocurrent" in Fig. 3.11

This block serves as a valuable tool for extracting key parameters of a commercial PV module, including light-generated current, the ideality factor and saturation current of the diode, and series and shunt resistances, all of which are essential for accurate PV module modeling. These parameters are displayed in Fig. 3.17 (highlighted by the blue rectangle) [51].

MATLAB Script-Based Modeling

In this section, the modeling of the PV module is performed based on the single-diode model using a MATLAB script, which is provided in Appendix A. Figure 3.18 presents the pseudo-code of the MATLAB script used to implement the mathematical model of the

3.2 Modeling of a Standalone Solar Photovoltaic System

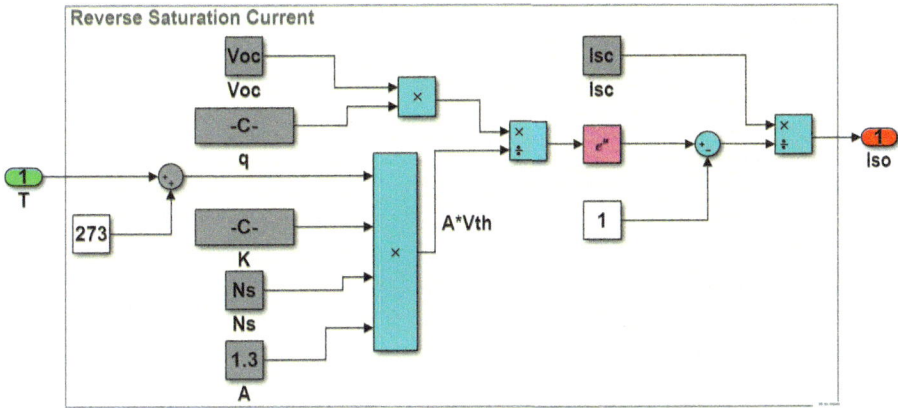

Fig. 3.13 Photocurrent implementation using Eq. (3.5). This represents the subsystem named "subsystem of photocurrent" in Fig. 3.11

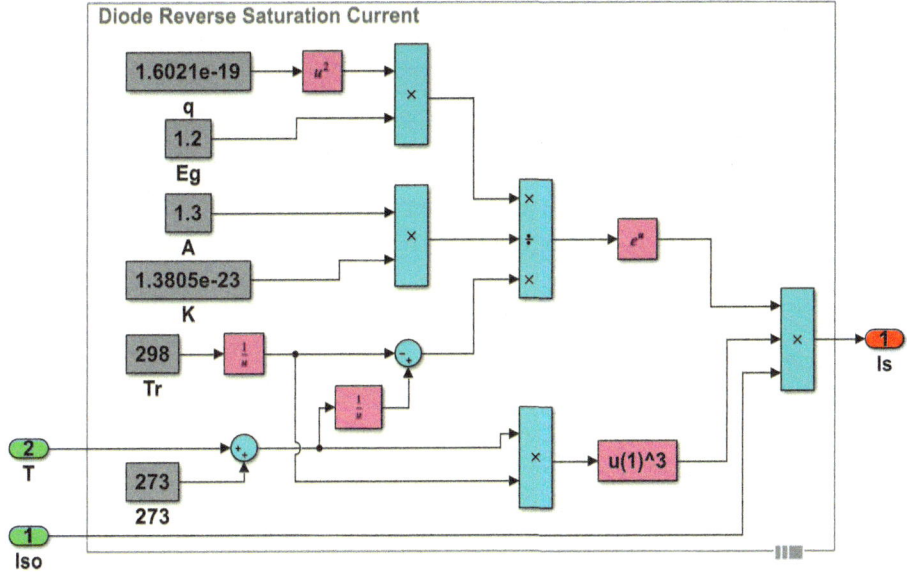

Fig. 3.14 Diode reverse saturation current implementation using Eq. (3.6). This subsystem is referred to as the "subsystem of diode reverse saturation current" in Fig. 3.11

PV module. It is important to note that the script employs the Newton-Raphson method for solving the nonlinear equations involved [21].

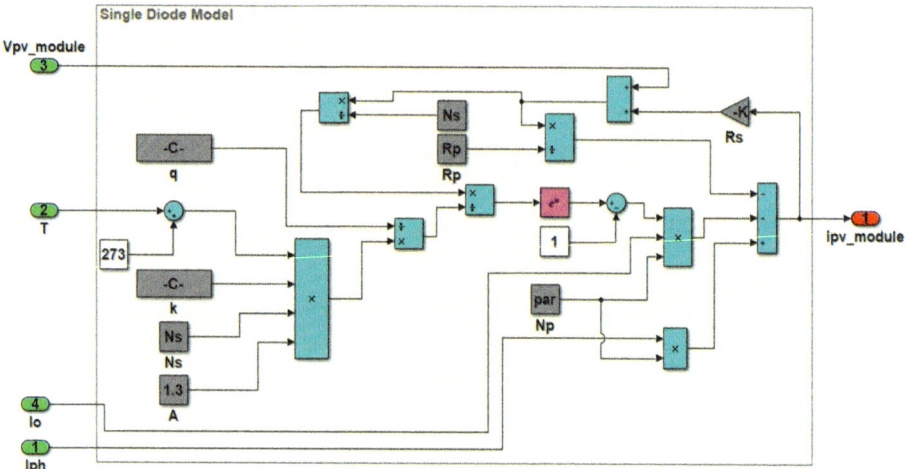

Fig. 3.15 Single-diode model implementation using Eq. (3.9). This subsystem is labeled "subsystem of single diode model" in Fig. 3.11

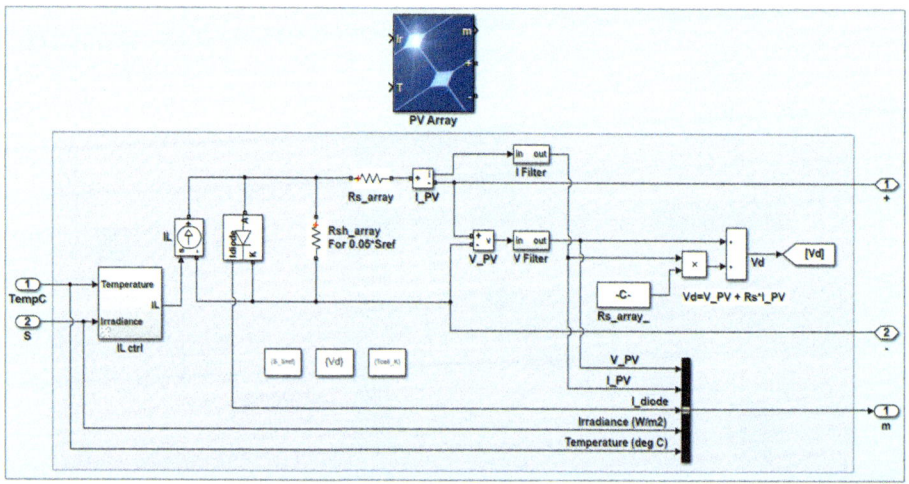

Fig. 3.16 Schematic diagram of the PV array block and its components

Proteus Platform-Based Modeling

Figure 3.19 illustrates the model of the PV module employed in this study, based on the five-parameter (single-diode) model using the Proteus platform. The equivalent circuit of the PV module consists of a current source, a shunt (parallel) resistor, a series resistor, and a diode, all integrated through a modified SPICE script [22, 23]. Figure 3.20 shows the

3.2 Modeling of a Standalone Solar Photovoltaic System

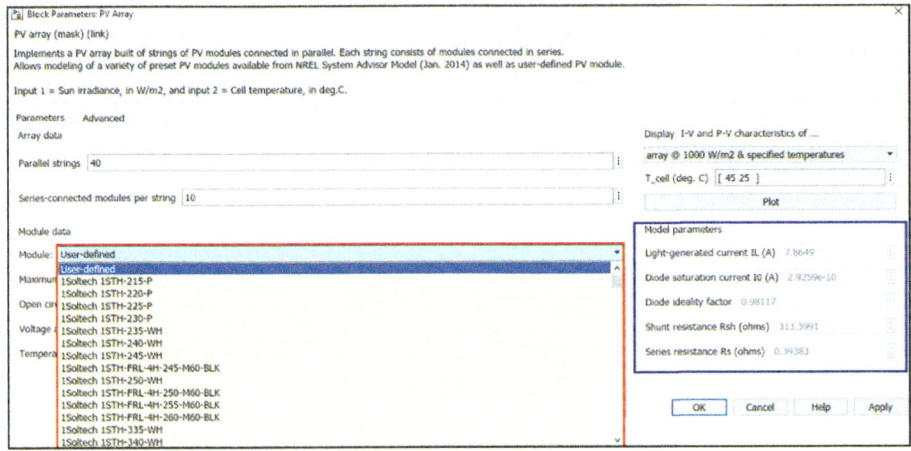

Fig. 3.17 Key parameters of the PV array block

```
1     %--------Characterization Of Photovoltaic Panel Using Single Diode Model--%
2     %------------------- Based on Newton-Raphson Method ----------------%
3
4 -   it=input('Enter the number of curves you aim to plot: 1,2...N: ');
5 -   for N=1:it
6     %-----------------------------------------------------------------%
7     %PV Module DATA
8     %-----------------------------------------------------------------%
9 -   T=input('Enter the value of T en °C: ');           % Temperature of the celle en °C
10 -  G=input('Enter the value of solar irradiance en °W/m2: '); % Solar Irradiance en W/m2
11 -  ai=0.102/100;                      % Current Tmperature coefficient(ki)
12 -  av=-0.36099/100;                   % Voltage Tmperature coefficient(kv)
13 -  Isc_r=7.84;                        % Short-circuit current
14 -  Voc_r=36.3;                        % Opent circuit voltage
15 -  Vm=29;                             % Maximum voltage @ STC
16 -  Im=7.35;                           % Maximum current @ STC
17 -  Pm=213.15;                         % Maximum power   @ STC
18 -  Ns=60;                             % nomber of Cells
19 -  n=0.98117;                         % Diode ideality factor
20    %-----------------------------------------------------------------%
```

Fig. 3.18 Pseudo-code of the MATLAB script for implementing the mathematical model of the PV module

circuit setup used to obtain the performance characteristics of the 1Soltech 1STH-215-P PV module, where a DC voltage unit is used as a variable load.

Comparison of the Four Methods

A comparative analysis was conducted on the PV module characteristic plots obtained from four different simulation methods: Simulink, the PV Array Block, MATLAB Script, and the Proteus implementation. The goal was to evaluate which method provided the most accurate simulation of the PV panel's behavior.

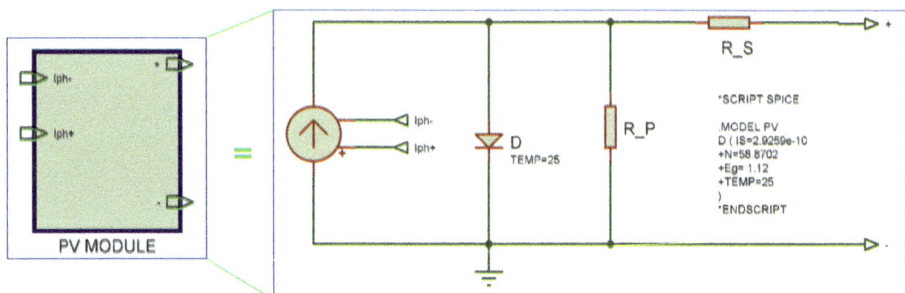

Fig. 3.19 Block diagram of the PV module and its circuit components within the Proteus environment

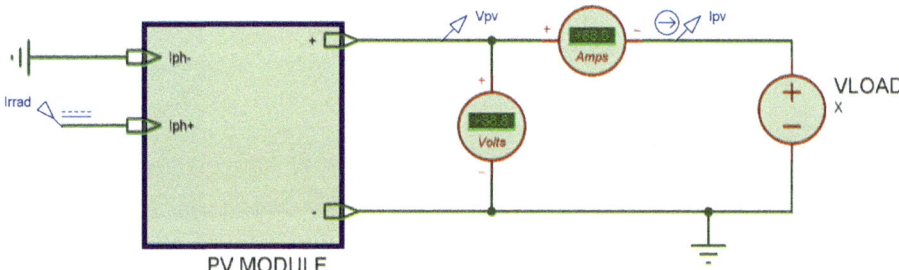

Fig. 3.20 Circuit configuration used for generating the characteristic curves of the PV module in the Proteus environment

The power-voltage (P–V) and current-voltage (I–V) characteristics of the 1Soltech 1STH-215-P PV module under standard test conditions (STC: 1 kW/m² irradiance and 25 °C) are displayed in Fig. 3.21. It can be observed that all the methods produced similar results, with only slight variations, which are highlighted in the zoomed-in sections of the graphs. However, the Proteus and PV Array Block methods demonstrated superior performance in terms of accuracy and reliability when compared to the other techniques. This comparison underscores the robustness of the Proteus platform, particularly for modeling and simulating PV modules, offering an efficient and reliable alternative to more traditional methods such as MATLAB scripts or Simulink.

The parameters of the 1Soltech 1STH-215-P PV module used in this study are reported in Table 3.2.

3.2.1.7 Effect of Meteorological Parameters on PV Module Performance

The performance of a PV module or array can be significantly influenced by various meteorological factors such as solar insolation, ambient temperature, partial shading, humidity,

3.2 Modeling of a Standalone Solar Photovoltaic System

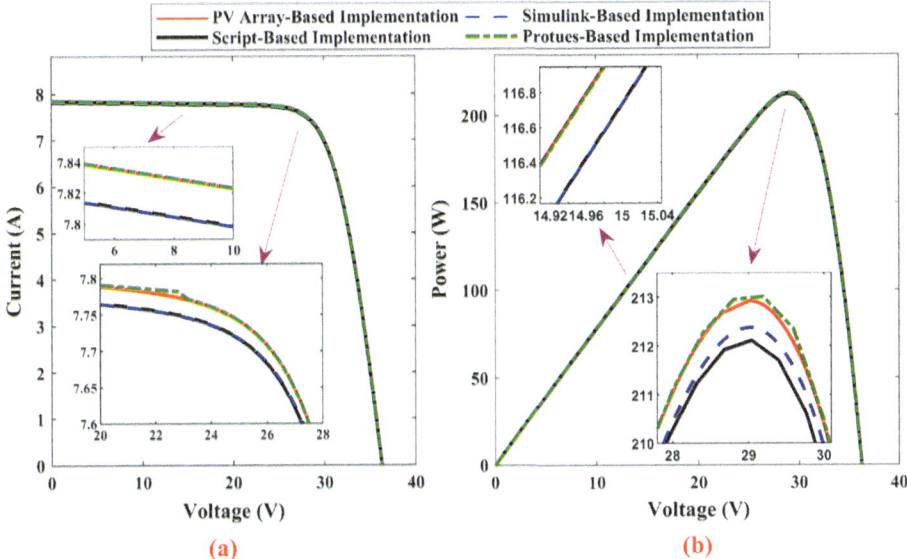

Fig. 3.21 Characteristics of the PV module under STC conditions: **a** I–V curve and **b** P–V curve

Table 3.2 Parameters of the 1Soltech 1STH-215-P PV module under STC

Parameters	Value
Maximum power (p_{mpp})	213.15 W
Voltage at MPP (V_{mpp})	29 V
Current at MPP (I_{mpp})	7.35 A
Open circuit voltage (V_{OC})	36.3 V
Short circuit current (I_{sc})	7.84 A
Temperature coefficient of (V_{OC})	$-0.36099\%/°C$
Temperature coefficient of (I_{sc})	$0.102\%/°C$
Diode saturation current (I_{sc})	2.9259×10^{-10} A
Shunt resistor (R_P)	313.3991 Ω
Series resistor (R_S)	0.39383 Ω
Ideality factor (A)	0.98117

and wind speed. However, based on Eqs. (3.5) to (3.9), it is evident that solar intensity and ambient temperature have the most profound impact on the physical behavior of the PV module/array. Consequently, the effects of solar intensity, partial shading, and ambient temperature on the performance parameters of the PV module, as presented in Table 3.2, were analyzed using MATLAB/Simulink.

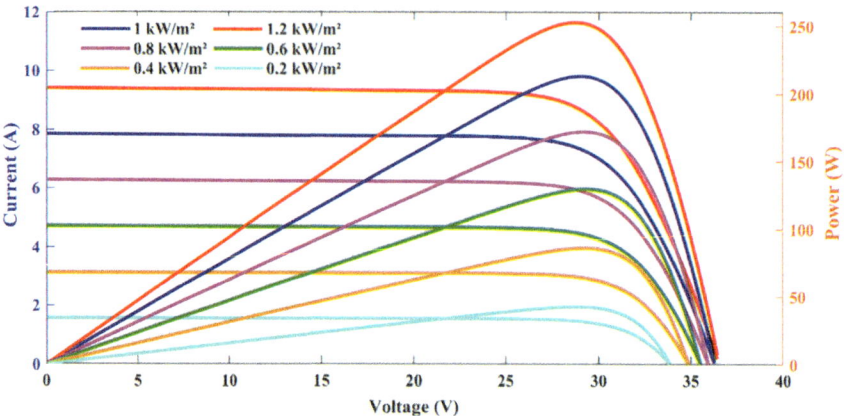

Fig. 3.22 P–V and I–V curves at varying solar intensity levels and a fixed temperature of 25 °C

Solar Intensity Effect

Figure 3.22 illustrates the impact of solar intensity on the P–V (power-voltage) and I–V (current-voltage) characteristics of the PV module at a fixed temperature of 25 °C. It is clear that the current and power output of the PV module are highly dependent on solar intensity levels. As the solar intensity increases, both the current and MPP of the module rise. However, the voltage of the PV module shows only minor variations with changes in solar intensity.

Ambient Temperature Effect

Figure 3.23 shows the effect of ambient temperature on the P–V and I–V characteristics of the PV module under a constant solar intensity of 1 kW/m^2. From these curves, it is evident that temperature significantly affects both the power and voltage of the PV module, while the current remains almost unchanged. As the temperature increases, both the power and voltage decrease. Additionally, Fig. 3.24 highlights a noticeable decline in the MPP of the PV module with rising temperatures at different levels of solar radiation [22].

Partial Shading Effect

Partial shading (PS) occurs when different areas of the PV module/array receive varying levels of solar intensity, caused by predictable objects like trees, towers, or power lines, as well as unpredictable factors such as fallen leaves or bird droppings [24]. Figure 3.25 demonstrates the P–V and I–V characteristics of the PV module under various partial shading conditions, showing multiple maxima such as local and global maximum power points (LMPP and GMPP). Therefore, partial shading has a substantial impact on the output power of the PV module.

3.2 Modeling of a Standalone Solar Photovoltaic System

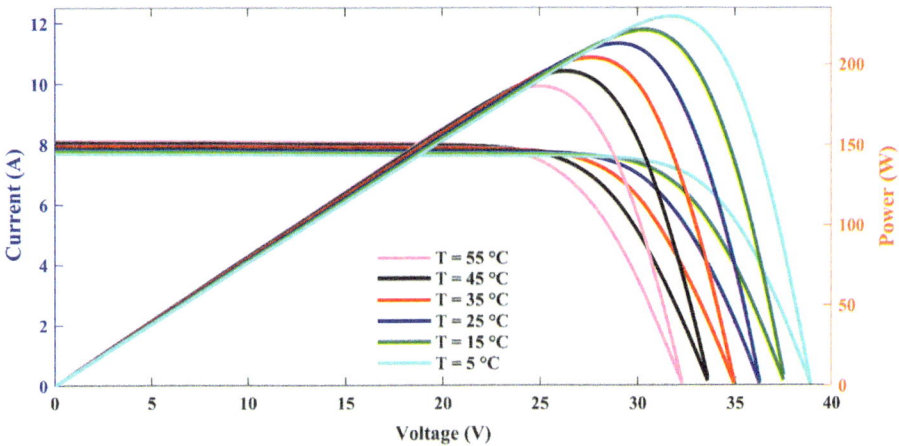

Fig. 3.23 P–V and I–V curves at various temperature levels with constant solar insolation of 1 kW/m²

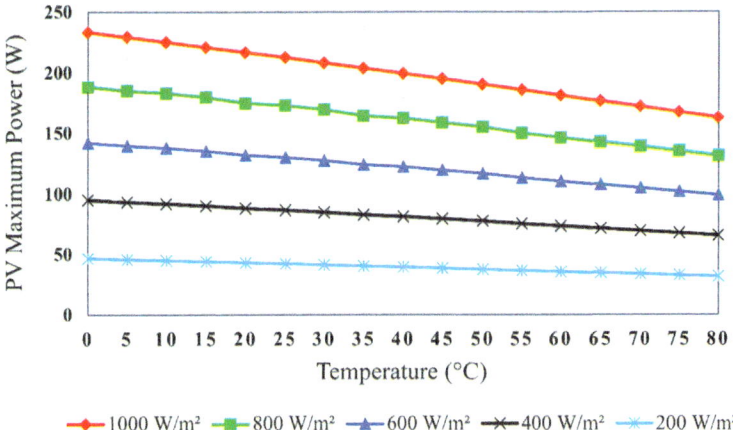

Fig. 3.24 MPP variations with increasing temperature under different solar radiation levels

3.2.1.8 Electrical Parameters Effect on PV Module Performance

In addition to meteorological parameters, which are external factors, there are also internal electrical parameters—such as shunt and series resistances, diode saturation current, and ideality factor—described in Eqs. (3.5) to (3.9). These internal parameters significantly affect the physical behavior of the PV module/array.

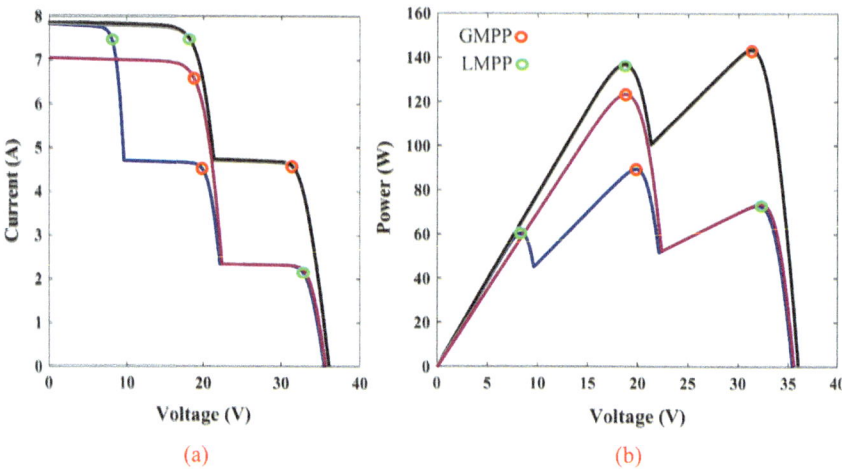

Fig. 3.25 **a** I–V and **b** P–V characteristics of the PV module under different partial shading conditions

Shunt Resistor Effect

The parallel or shunt resistance (*Rp*) represents any high-conductivity path across the p–n junction or along the edges of the PV cell [24, 25]. Figure 3.26 illustrates the effect of shunt resistance on the I–V and P–V output characteristics of the PV module. A low *Rp* value causes a reduction in current and MPP. Conversely, the higher the Rp value, the greater the current and MPP. Ideally, *Rp* should be as high as possible to maximize efficiency. In this analysis, the solar intensity and temperature are kept constant at 1 kW/m^2 and 25 °C.

Series Resistor Effect

The series resistance (*Rs*) accounts for the total resistance of the rear metal, finger contacts, bus bars, and the front and rear contacts of the PV module. Figure 3.27 depicts the impact of Rs on the I–V and P–V curves. It is clear that lower Rs values yield more ideal I–V and P–V characteristics, while higher Rs values cause a significant reduction in MPP. Additionally, Rs decreases linearly with increasing temperature, as reported in the literature [26–28]. The simulation maintains constant solar intensity and temperature at 1 kW/m^2 and 25 °C.

Diode Saturation Current Effect

Figure 3.28 displays the I–V and P–V characteristics of the PV module, considering different values of diode saturation current (*Is*) under constant insolation and temperature (1 kW/m^2 and 25 °C). As seen, an increase in Is results in a noticeable decrease in both open-circuit voltage (*Voc*) and MPP [29, 30].

3.2 Modeling of a Standalone Solar Photovoltaic System

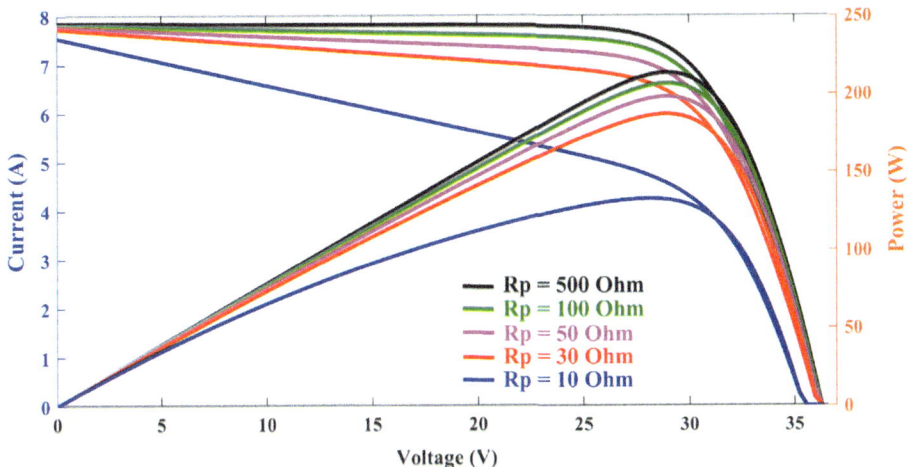

Fig. 3.26 P–V and I–V curves at varying shunt resistance values

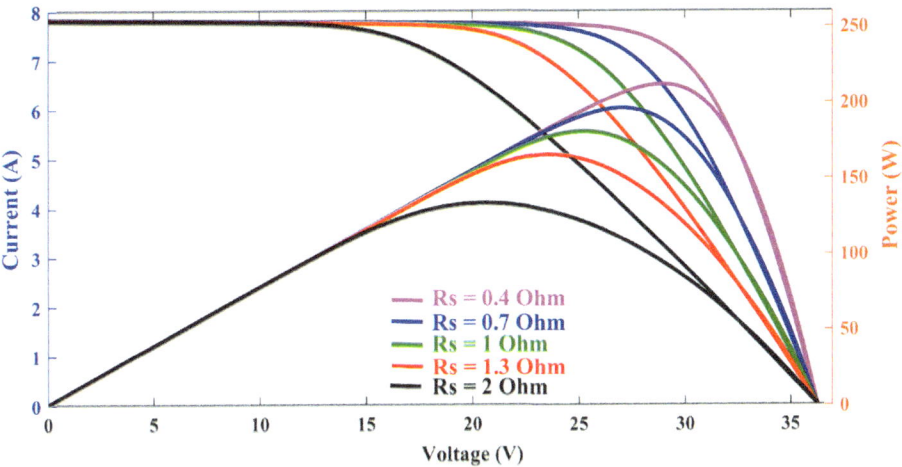

Fig. 3.27 P–V and I–V curves at different series resistance values

Ideality Factor Effect

The ideality factor (A) measures the deviation of a diode's behavior from the ideal diode equation. Its value generally ranges between 1 and 2 and depends on the PV technology used, as shown in Table 3.3 [26]. Figure 3.29 highlights how variations in the ideality factor influence the I–V and P–V curves, particularly the MPP region. The closer A is to 1, the more ideal the characteristics of the PV module, leading to a higher MPP.

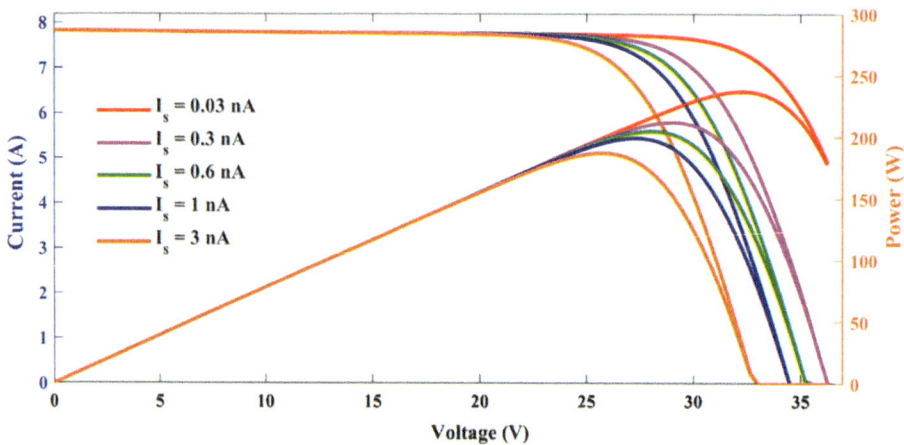

Fig. 3.28 P–V and I–V curves under various diode saturation current values

Table 3.3 Ideality factor for different PV technologies

PV technology	Si-poly	Si-mono	a-Si: H	CdTe	CIS	GaAs
A	1.30	1.2	1.8	1.5	1.5	1.3

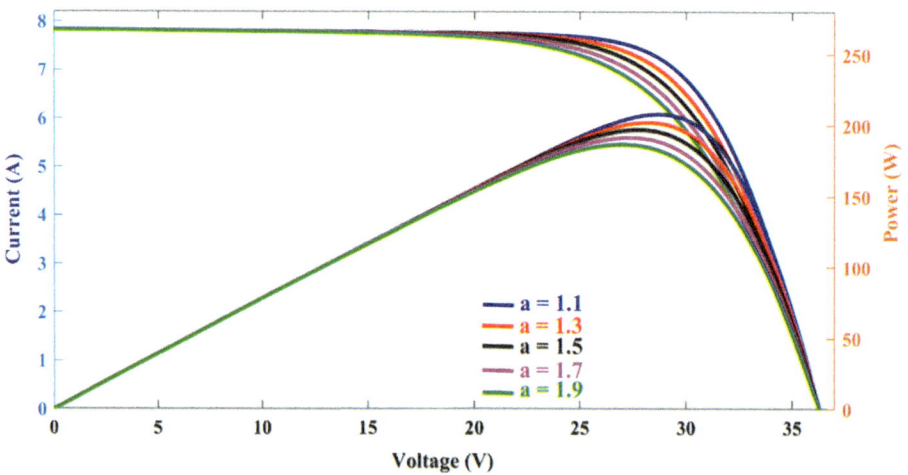

Fig. 3.29 P–V and I–V curves at different values of the ideality factor

3.2 Modeling of a Standalone Solar Photovoltaic System

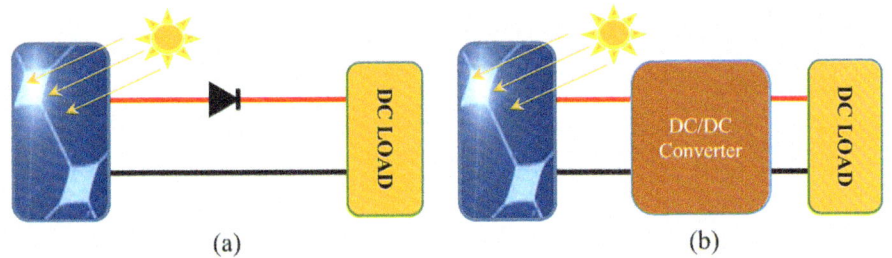

Fig. 3.30 Coupling modes of the PV module and load in a standalone PV system: **a** Direct connection, **b** Via a DC-DC converter

3.2.2 MPPT System Modeling

3.2.2.1 Direct Connection of a PV Module/Array and DC Load

When a PV module is directly connected to a DC load, as shown in Fig. 3.30a, the operating point (OP) is located at the intersection of the I–V curve of the PV module and the load line, as depicted in Fig. 3.31. This operating point often does not coincide with the maximum power point (MPP) of the PV module or array, and it varies with the value of the load resistance (R_L). For example, Fig. 3.31 shows that the power output at operating points A and B is significantly lower than at the MPP. This configuration is inefficient due to power losses, resulting in lower overall energy yield. To address this issue, a DC-DC power converter is required to adjust the electrical operating points of the PV module and the load, as shown in Fig. 3.30b.

3.2.2.2 Modeling and Control of DC-DC Converters Using Simulink Platform

DC-DC converters play a crucial role in maximizing energy conversion efficiency in PV systems. They force the PV module/array to operate at its MPP voltage, ensuring the efficient transfer of maximum power from the PV module/array to the load. In this thesis, different types of DC-DC converters such as boost, buck, interleaved boost (IB), and three-level boost (TLB) converters have been implemented and modeled using the MATLAB/Simulink platform. Circuitry modeling was chosen over other techniques, such as mathematical modeling, transfer functions, or state-space modeling, due to its simplicity and flexibility [31]. The parameters used in the simulation include: $C_{IN} = C_O = C_1 = C_2 = 100 \, \mu F$, $L = L_1 = L_2 = 3 \, mH$, PWM frequency (f) = 20 kHz, duty cycle (d) = 0.5, and $R_L = 10 \, \Omega$. These parameters are selected based on equations outlined in Chapter 2.

DC-DC Boost Converter

The boost converter, also known as a step-up converter, is one of the most commonly used DC-DC converters in PV systems due to its simplicity and ability to efficiently transfer maximum power from the PV module/array to the load with minimal energy losses [23].

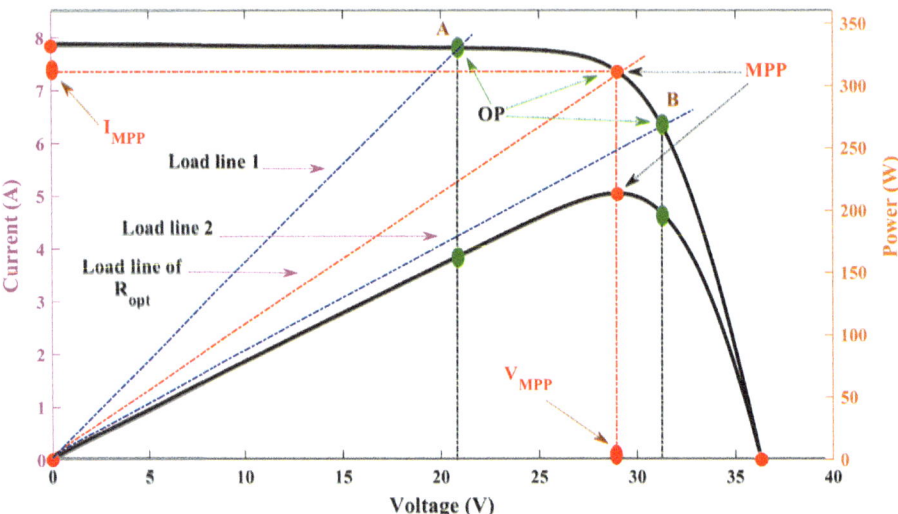

Fig. 3.31 Operating point (OP) on the I–V and P–V curves of a PV module with a DC load

Fig. 3.32 Schematic circuit of the DC-DC boost converter

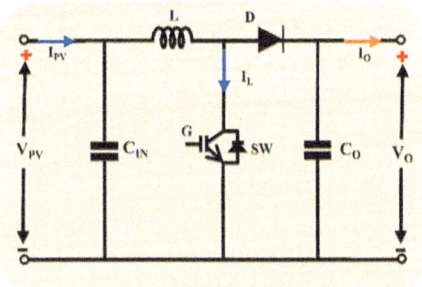

Figure 3.32 presents the schematic circuit of the DC-DC boost converter, and its electrical model implementation in Simulink is shown in Fig. 3.33. The input and output voltage and current waveforms are displayed in Fig. 3.34.

DC-DC Buck Converter

The buck converter, also known as a step-down converter, is widely used in PV systems where the voltage of the PV module needs to be stepped down to match the load or battery voltage [34, 35]. This is particularly useful in applications like solar-powered water pumps, battery charging stations, and voltage regulation systems [32, 33, 36]. The schematic circuit of the buck converter is shown in Fig. 3.35, and its circuit modeling and simulation results are presented in Figs. 3.36 and 3.37, respectively.

3.2 Modeling of a Standalone Solar Photovoltaic System

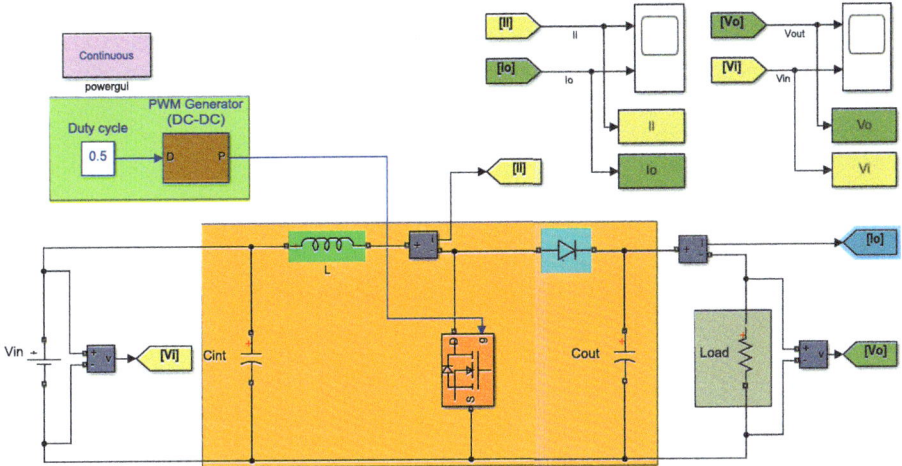

Fig. 3.33 Circuit modeling implementation of the boost converter using the Simulink platform

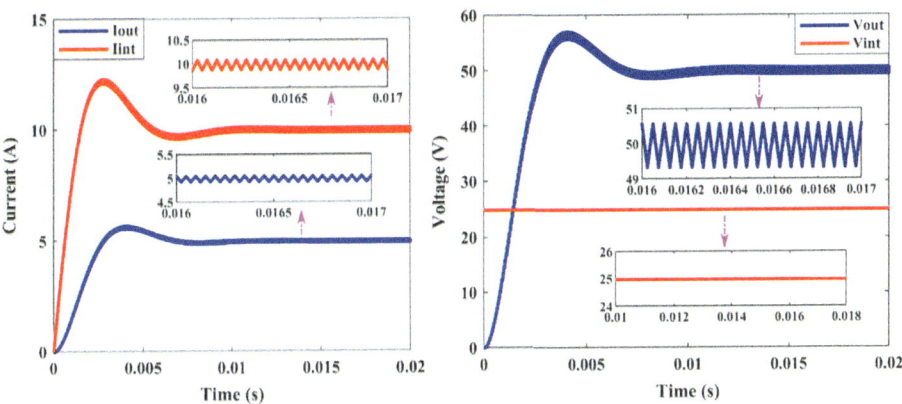

Fig. 3.34 Input and output voltage and current waveforms of the boost converter

Fig. 3.35 Schematic circuit of the buck converter

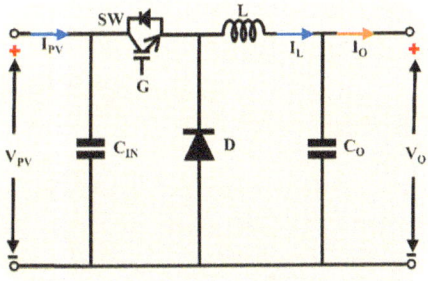

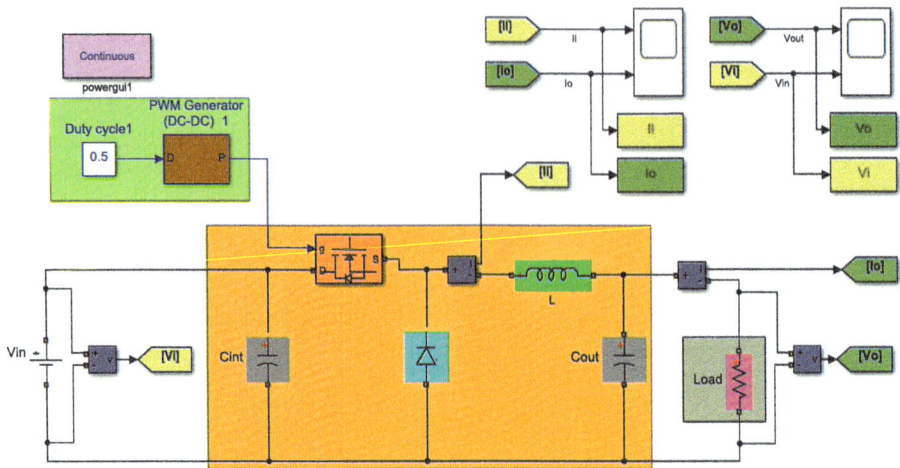

Fig. 3.36 Circuit modeling of the buck converter using the Simulink platform

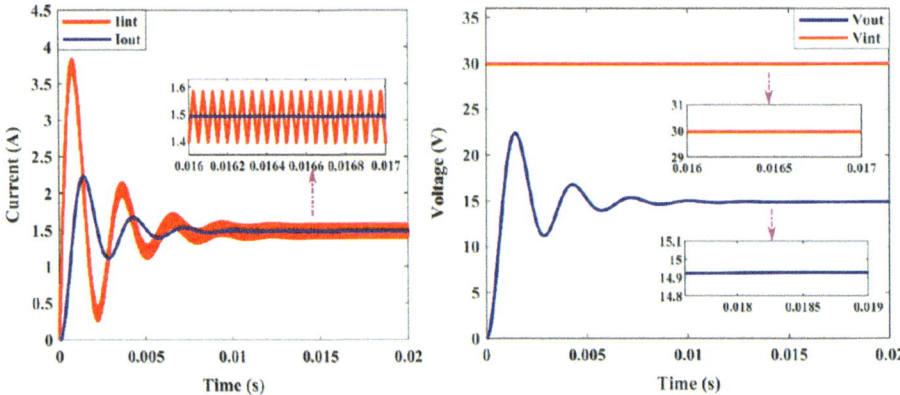

Fig. 3.37 Input and output voltage and current curves of the buck converter

DC-DC Interleaved Boost Converter

The interleaved boost (IB) converter, shown in Fig. 3.38, has gained considerable traction in photovoltaic applications due to its advantages such as lower current ripple, wider continuous current mode, reduced output voltage ripple, lower switching losses, and higher power conversion efficiency compared to conventional boost converters [37]. By splitting the current into two power paths, significant conductivity losses (I^2R) can be minimized [37]. The circuit modeling of the IB converter in Simulink is depicted in Fig. 3.39, with two inductors, two diodes, input/output capacitor filters, and two switches controlled by

3.2 Modeling of a Standalone Solar Photovoltaic System

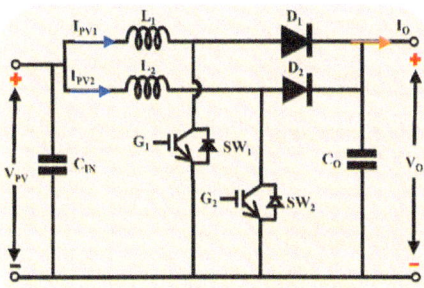

Fig. 3.38 Schematic circuit of the interleaved boost converter

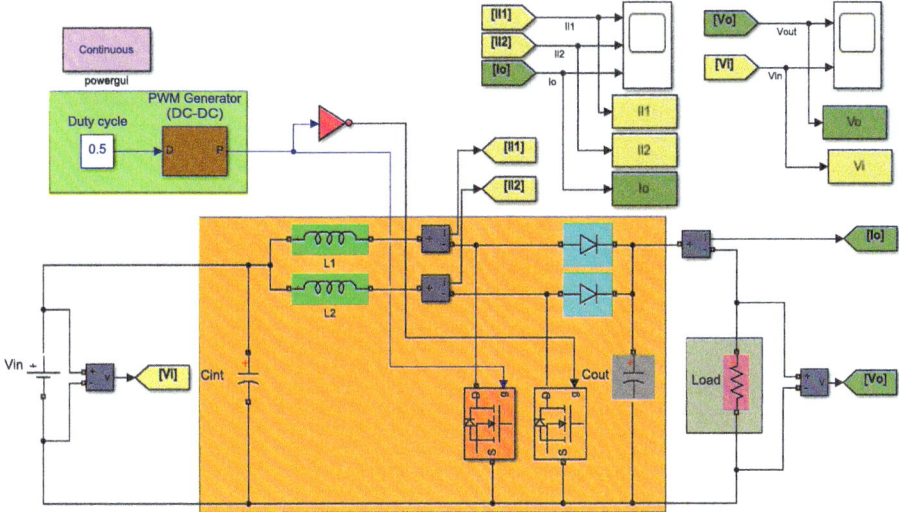

Fig. 3.39 Circuit modeling of the IB converter using the Simulink platform

a 180° phase-shifted PWM signal. Figure 3.40 shows the input and output voltage and current waveforms from the simulation.

DC-DC Three-Level Boost Converter

The three-level boost (TLB) converter has become increasingly important in modern power electronics due to its application in high-power systems where high voltage ratings are required, such as electric vehicle charging stations and renewable energy systems, including photovoltaic and wind energy [38–40]. Compared to conventional boost converters, the TLB offers benefits like reduced voltage stress on output capacitors, lower switching losses, and reduced inductor current ripple. Figures 3.41 and 3.42 show the

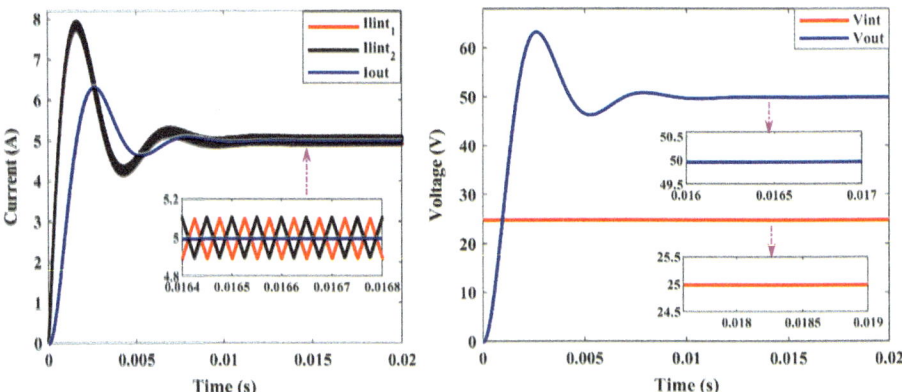

Fig. 3.40 Input and output waveforms of the interleaved boost converter

Fig. 3.41 Schematic circuit of the three-level boost converter

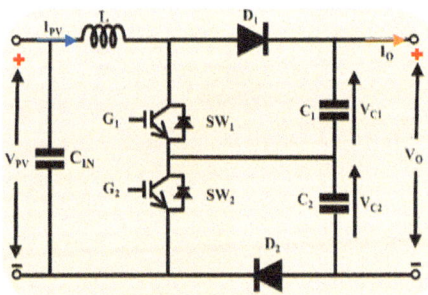

schematic and modeling circuits of the TLB converter, while Fig. 3.43 displays the input and output voltage and current waveforms.

3.2.2.3 Requirement for MPPT Algorithm

As discussed in Sect. 3.2.2.1, directly connecting a PV module or array to a DC load results in significant power losses and reduced energy yield. To mitigate this, a DC-DC conversion stage is typically introduced to match the load impedance with that of the PV module or array, allowing the operating point of the system to align with the MPP of the PV module or array. However, the MPP is not constant and depends on varying atmospheric conditions such as solar insolation and ambient temperature. To ensure that the system continuously operates at the MPP, a control mechanism is needed to manage the DC-DC conversion stage. This control can be achieved using an MPPT algorithm [41, 42].

3.2 Modeling of a Standalone Solar Photovoltaic System

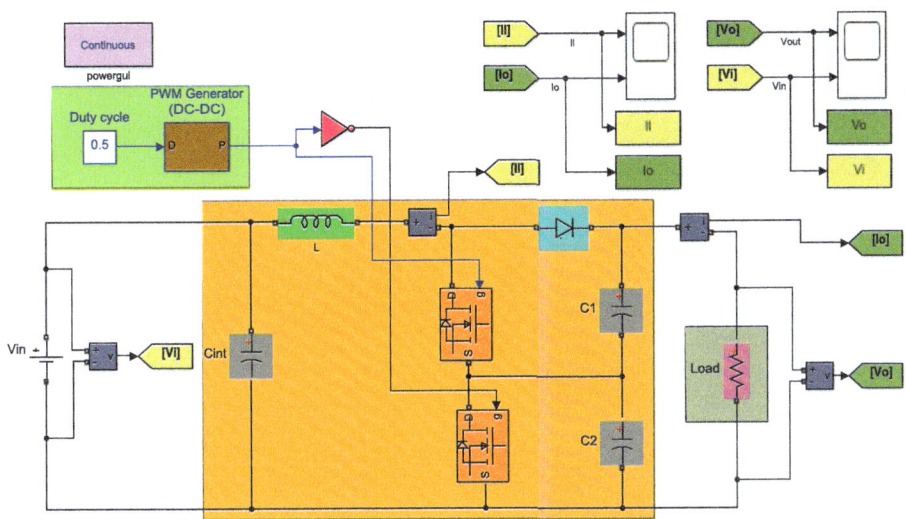

Fig. 3.42 Circuit modeling of the three-level boost converter using the Simulink platform

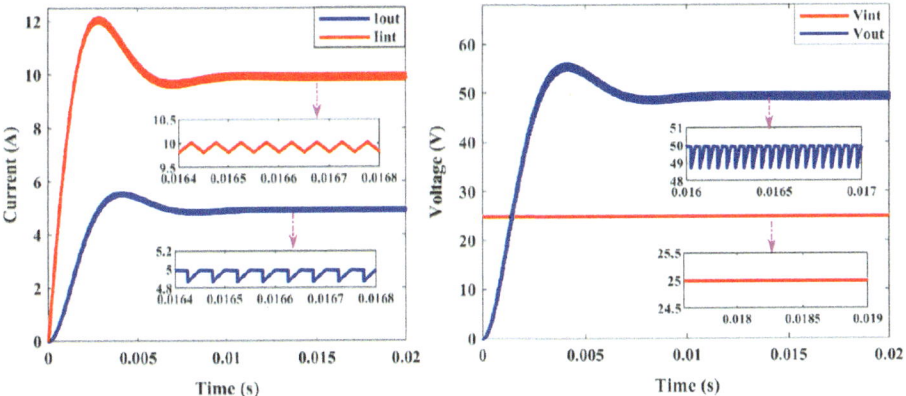

Fig. 3.43 Input and output voltage and current waveforms of the three-level boost converter

3.2.3 MATLAB/Simulink Software-Based Modeling

The complete standalone solar PV system is modeled using MATLAB/Simulink, as shown in Fig. 3.44. The system consists of three main components: the source, the interface, and the load.

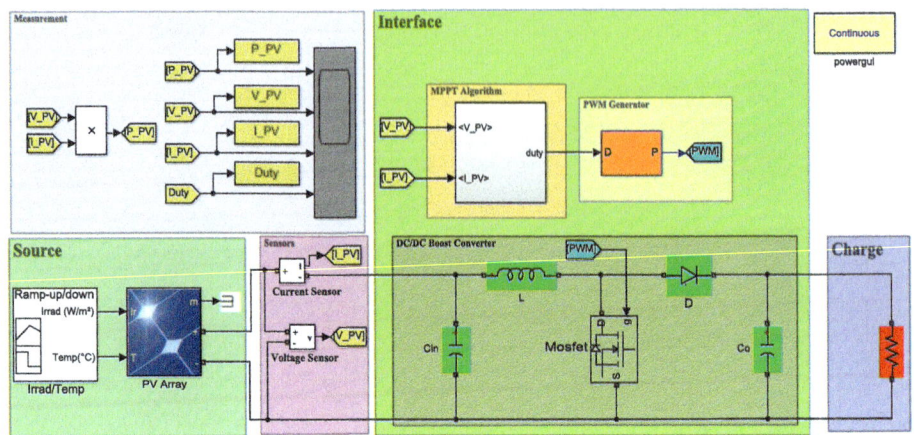

Fig. 3.44 MATLAB/Simulink model of the overall standalone solar PV system

- The source part includes the photovoltaic module or array.
- The interface part consists of the MPPT algorithm, a DC-DC boost converter, and a PWM generator.
- The load part is represented by a resistive load.

Additionally, two sensors are integrated for current and voltage measurement, and a powergui block is included for simulating Simscape Power Systems models.

3.2.4 Proteus Software Based Modeling

The use of Proteus software for PV system implementation has gained popularity due to its simplicity and extensive libraries of electronic components [43]. One of its key advantages is that it allows users to control the PV system using low-cost microcontrollers like Arduino boards. This makes it possible to achieve simulation results similar to those obtained through experimental validation [44].

Figure 3.45 illustrates the standalone solar PV system modeled in the Proteus environment. The system includes essential components such as the PV module, boost converter, resistive load, INA169 analog DC current sensor, B25 voltage sensor, Arduino Uno board with ATmega328 microcontroller for MPPT implementation, TC4420 driver, and a 16-bit LCD display [46].

3.3 Simulation of a Standalone Solar PV System with Conventional ...

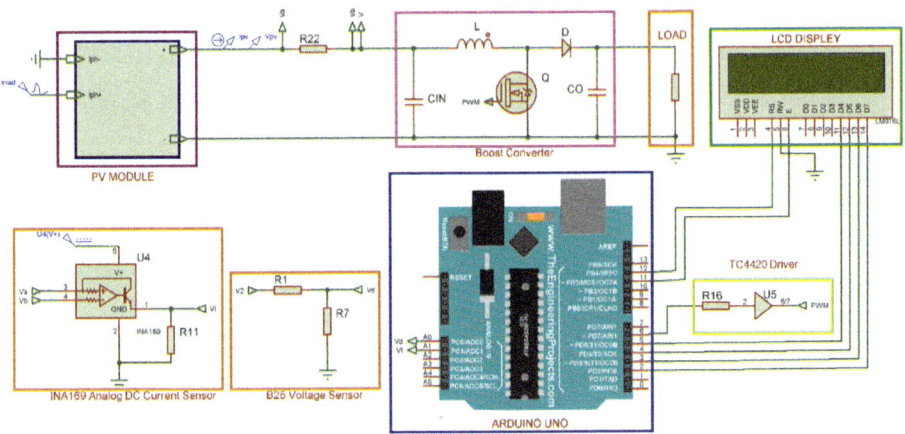

Fig. 3.45 Proteus software model of the overall standalone solar PV system

3.3 Simulation of a Standalone Solar PV System with Conventional MPPT Algorithms

In this section, the overall standalone solar PV system is simulated using conventional MPPT algorithms like P&O and INC techniques. The simulations are performed using two software platforms: MATLAB/Simulink and Proteus. The parameters used in the simulation for both platforms are outlined in Table 3.4.

3.3.1 P&O MPPT Algorithm-Based Implementation

The P&O algorithm is one of the most widely used MPPT techniques due to its simplicity and ease of implementation [47–49]. However, it has some limitations, such as power ripples caused by oscillations around the MPP in steady-state conditions, and slow response time in situations where there is a rapid change in solar intensity or temperature.

The flowchart of the P&O MPPT algorithm is presented in Fig. 3.46, and the algorithm is implemented in the Simulink platform using an M-file code in an embedded MATLAB function, as shown in Fig. 3.47. The full MATLAB script for the P&O MPPT algorithm is provided in Appendix B.

3.3.1.1 Simulation Results Based on MATLAB/Simulink Platform Implementation

The simulation results for the complete standalone solar PV system employing the P&O MPPT technique in the MATLAB/Simulink environment are presented in Figs. 3.49, 3.50, 3.51, 3.52 and 3.53. These simulations were performed under rapidly changing solar

Table 3.4 Parameters used in the implementation of the standalone solar PV system in MATLAB/Simulink and Proteus

Parameters	MATLAB/simulink environment	Proteus environment
	Value	
PV module		
Maximum power (P_{MPP}) (W)	213.15	17
Voltage at MPP (V_{MPP}) (V)	29	18.8
Current at MPP (I_{MPP}) (A)	7.35	0.9
Open circuit voltage (V_{OC}) (V)	36.3	22.5
Short circuit current (I_{SC}) (A)	7.84	1.01
Temperature coefficient of V_{OC} %/°C	−0.36099	−0.35
Temperature coefficient of I_{SC} %/°C	0.102	0.043
DC–DC boost converter		
Input capacitor (μF)	47	220
Output capacitor (μF)	470	470
Inductance (mH)	1	20
Switching frequency (kHz)	10	1
Charge		
Resistive load (Ω)	30	70

insolation, as shown in Fig. 3.48, while maintaining a fixed temperature of 25 °C. The parameters used for the entire standalone solar PV system are listed in Table 3.4.

In Figs. 3.49, 3.50, and 3.51, the PV output power, duty cycle, and PV voltage and current waveforms are shown for the P&O MPPT method using two different perturbation step sizes: 0.01 and 0.005. It is evident from the PV output power curves in Fig. 3.49 that the P&O algorithm successfully tracks the MPP with both step sizes, though the tracking performance differs.

When using a larger step size of 0.01 in the duty cycle perturbation (Fig. 3.50), the P&O algorithm quickly reaches the MPP. However, significant oscillations in the PV output power occur, leading to increased power losses. In contrast, when using a smaller perturbation step size of 0.005, the oscillations around the MPP are reduced, but the convergence speed to the MPP is slower, as seen in Fig. 3.49.

The PV output power and duty cycle versus PV output voltage curves, obtained with the P&O MPPT algorithm, are depicted in Figs. 3.52 and 3.53, respectively. These figures clearly illustrate the tracking process of the P&O MPPT method and the associated oscillation issues it faces.

3.3 Simulation of a Standalone Solar PV System with Conventional ...

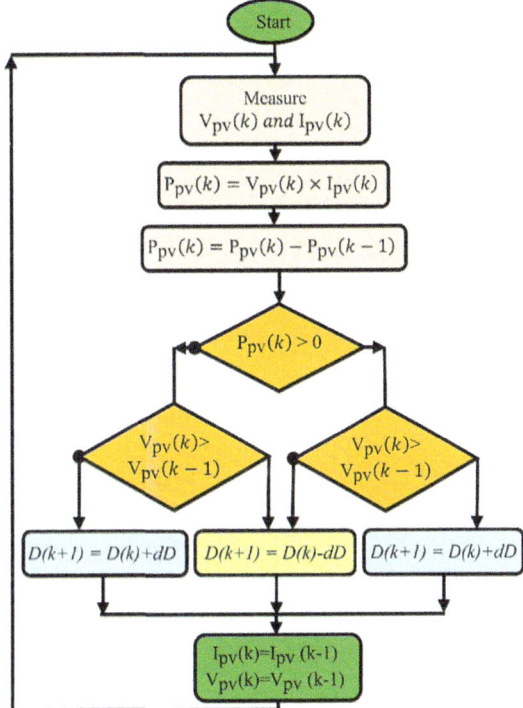

Fig. 3.46 Diagram of the P&O MPPT algorithm

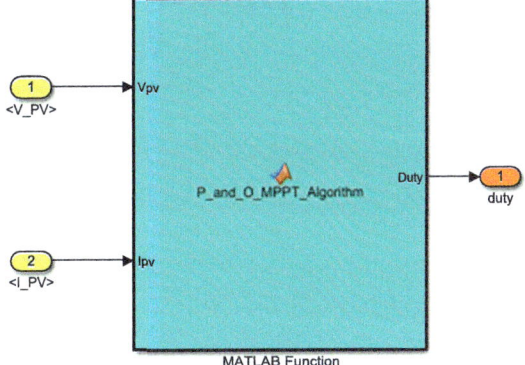

Fig. 3.47 Simulink implementation of the P&O MPPT algorithm using M-file code in an embedded MATLAB function

3.3.1.2 Simulation Results Using Proteus Platform Implementation

The implementation and simulation of the complete standalone PV system using the Proteus environment are shown in Fig. 3.54. In this setup, the Arduino UNO board is used to implement the P&O MPPT algorithm (the code for which is provided in Appendix C). The simulation results for the PV output power waveforms, tracked by the P&O MPPT method with two perturbation step sizes (0.2 and 0.5), are displayed in Fig. 3.55.

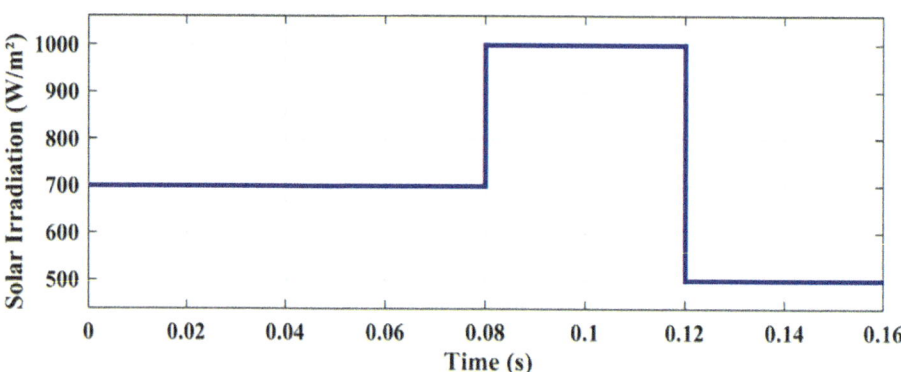

Fig. 3.48 Profile of solar irradiation variation

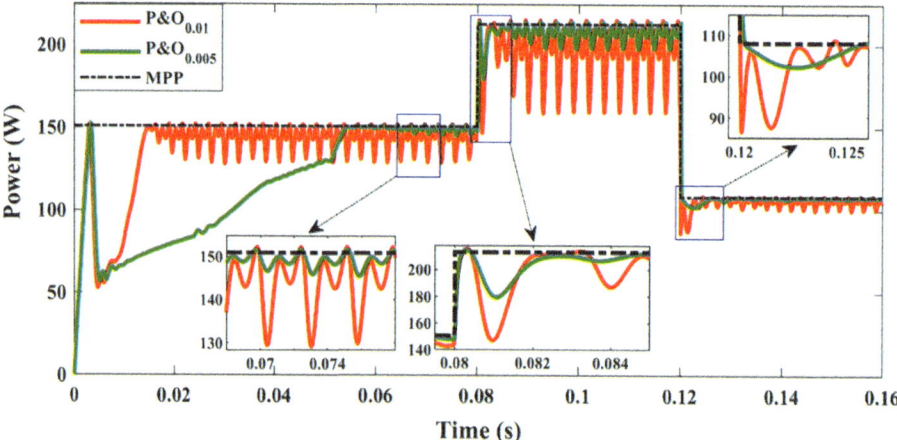

Fig. 3.49 PV output power curves obtained using the P&O MPPT algorithm in the Simulink environment

The results demonstrate the effect of different step sizes on the performance of the P&O algorithm. As seen, a larger step size results in faster convergence but causes larger oscillations around the MPP, while a smaller step size reduces oscillations but slows down the tracking speed.

3.3.2 INC MPPT Algorithm-Based Implementation

The INC MPPT strategy has been widely adopted in various studies [45, 50] to overcome the drawbacks associated with the conventional P&O MPPT scheme. The INC MPPT

3.3 Simulation of a Standalone Solar PV System with Conventional …

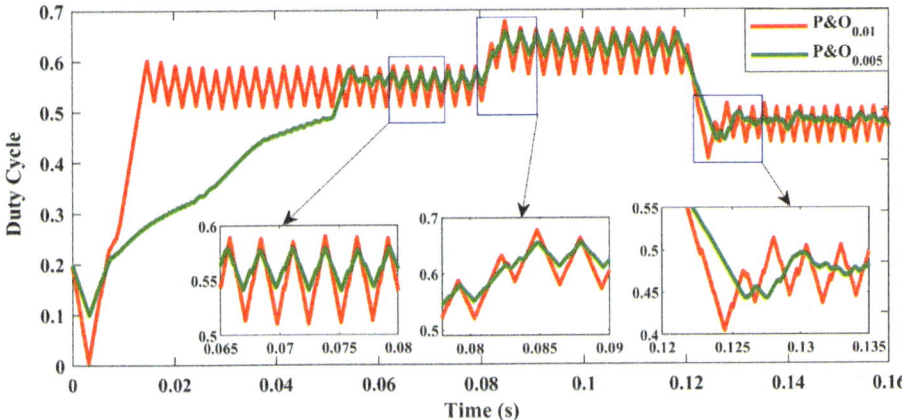

Fig. 3.50 Duty cycle curves obtained using the P&O MPPT algorithm in the Simulink environment

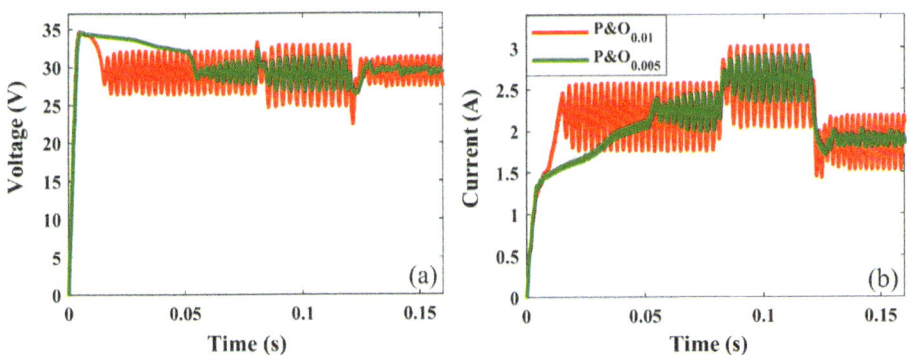

Fig. 3.51 PV voltage and current curves obtained using the P&O MPPT algorithm in the Simulink environment

method tracks the MPP by comparing the instantaneous conductance $\left(-I_{pv}/V_{pv}\right)$ with the incremental conductance $\left(\Delta I_{pv}/\Delta V_{pv}\right)$. At the MPP, the incremental conductance equals zero, while it is less than zero when the MPP is on the right side of the curve, and greater than zero when the MPP is on the left side [51]. The diagram of the INC MPPT algorithm is illustrated in Fig. 3.56, and its implementation using an embedded MATLAB function M-file code in the Simulink platform is shown in Fig. 3.57. The complete M-file code of the INC MPPT algorithm can be found in Appendix D.

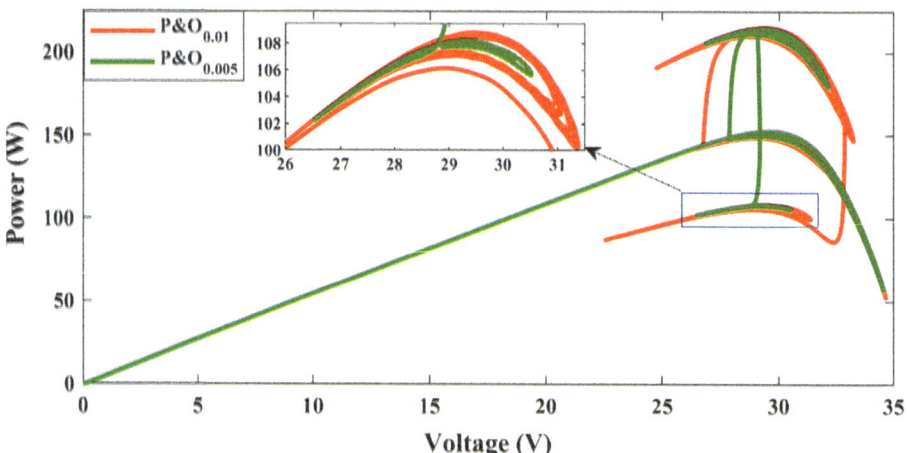

Fig. 3.52 PV power versus voltage curves obtained using the P&O MPPT method in the Simulink environment

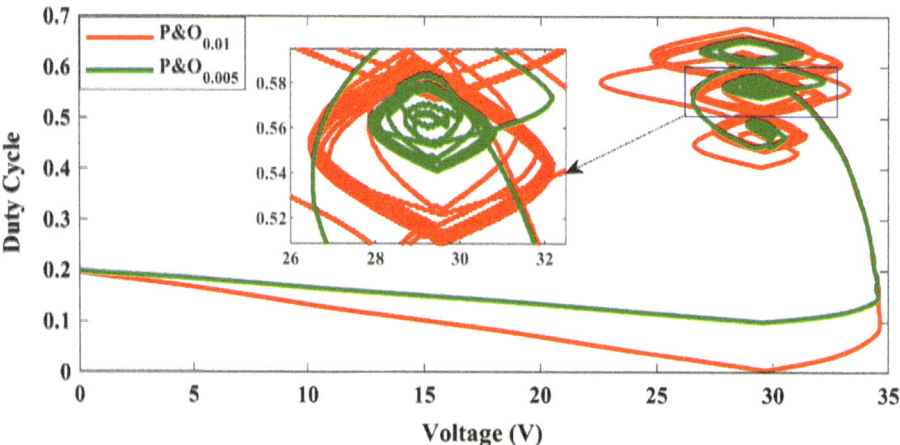

Fig. 3.53 Duty cycle versus voltage curves obtained using the P&O MPPT method in the Simulink environment

3.3.2.1 Simulation Results Using the MATLAB/Simulink Platform Implementation

The simulation results for the complete standalone solar PV system using the INC MPPT method on the Simulink platform are shown in Figs. 3.58, 3.59, 3.60, 3.61 and 3.62. These simulations were conducted under the same conditions used for the P&O MPPT algorithm. The figures illustrate the PV output power, duty cycle, and PV output voltage

3.3 Simulation of a Standalone Solar PV System with Conventional …

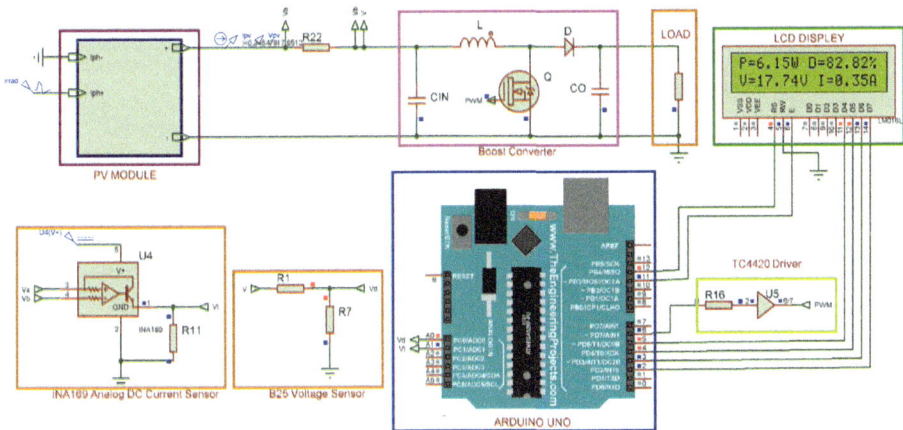

Fig. 3.54 Overall implementation of the standalone PV system using the P&O MPPT method in the Proteus environment

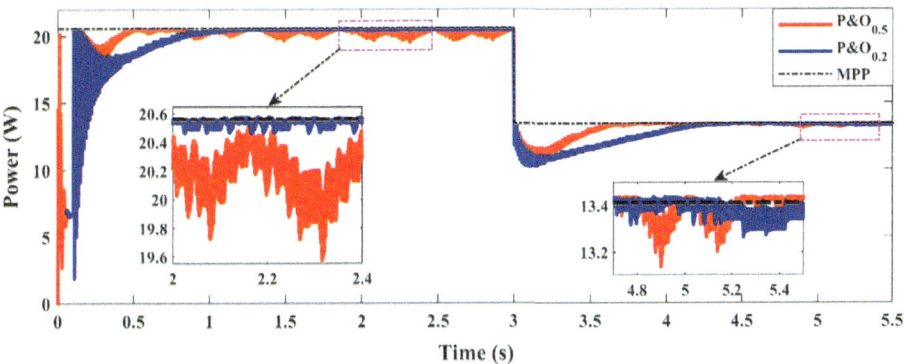

Fig. 3.55 PV output power waveforms obtained using the P&O MPPT method in the Proteus environment

and current waveforms for the INC MPPT method using two different perturbation step sizes: 0.01 and 0.005.

Figures 3.58, 3.59 and 3.60 display the output waveforms for PV power, duty cycle, and PV voltage/current, respectively. It is evident from the results that the INC MPPT method encounters similar issues as the P&O MPPT method, including power oscillations around the MPP.

Additionally, Figs. 3.61 and 3.62 depict the PV power versus voltage and duty cycle versus voltage curves for the INC MPPT method. These results highlight the fluctuation and instability issues in the INC method when attempting to accurately track the MPP.

Fig. 3.56 Diagram of the INC MPPT algorithm

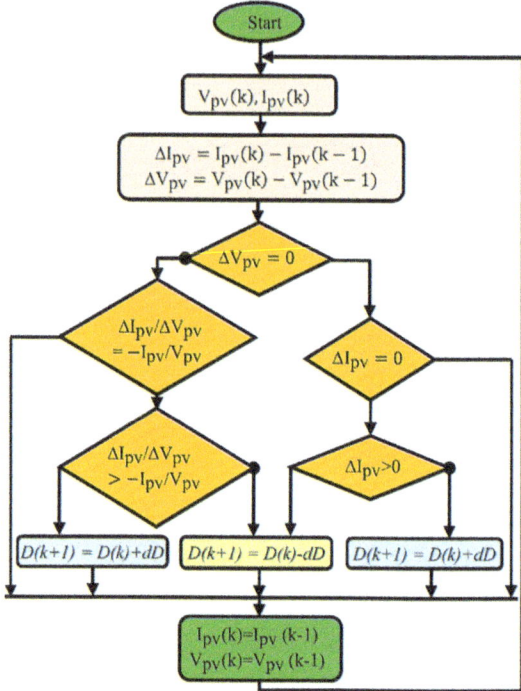

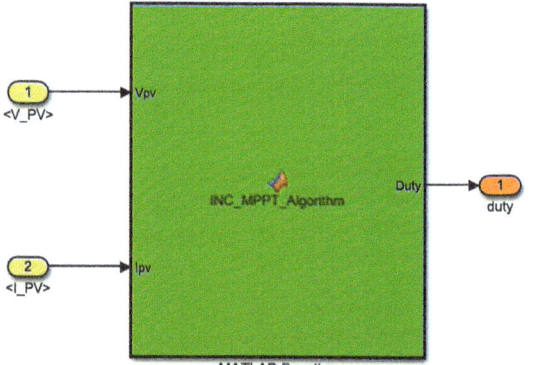

Fig. 3.57 Implementation of the INC MPPT algorithm using M-file code in an embedded MATLAB function

3.3.2.2 Simulation Results Using the Proteus Platform Implementation

The simulation results for the PV output power obtained using the INC MPPT technique in the Proteus environment are shown in Fig. 3.63. The INC MPPT method was implemented with two perturbation step sizes, 0.2 and 0.5 (the code for this implementation can be found in Appendix E). It is important to note that the same standalone PV system, previously shown in Fig. 3.63, was used for the implementation of the INC MPPT technique.

3.3 Simulation of a Standalone Solar PV System with Conventional … 87

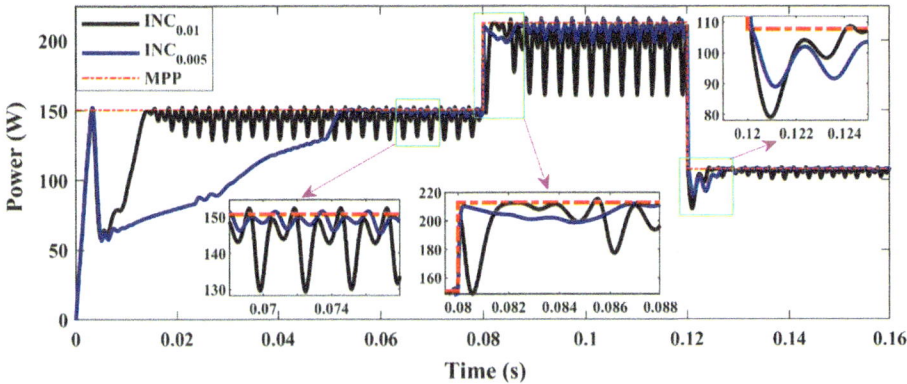

Fig. 3.58 PV module output power tracking curves obtained using the INC MPPT method

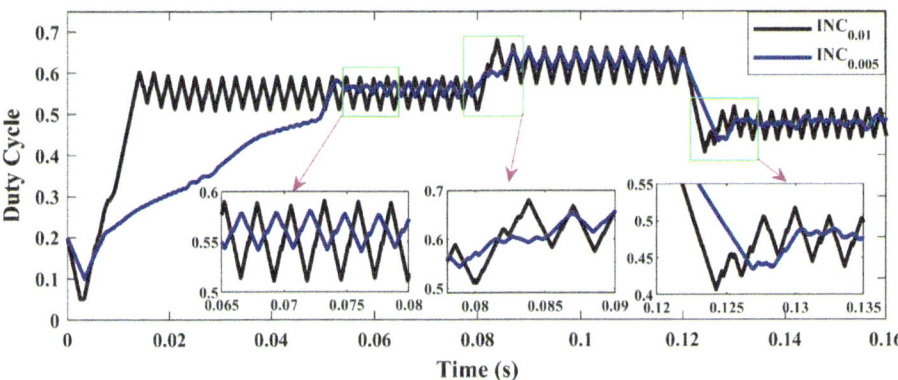

Fig. 3.59 Duty cycle simulation curves obtained using the INC MPPT method

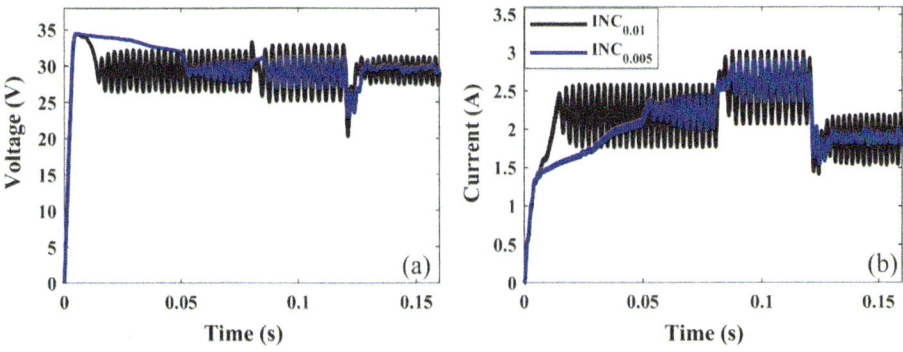

Fig. 3.60 PV voltage and current simulation curves obtained using the INC MPPT method

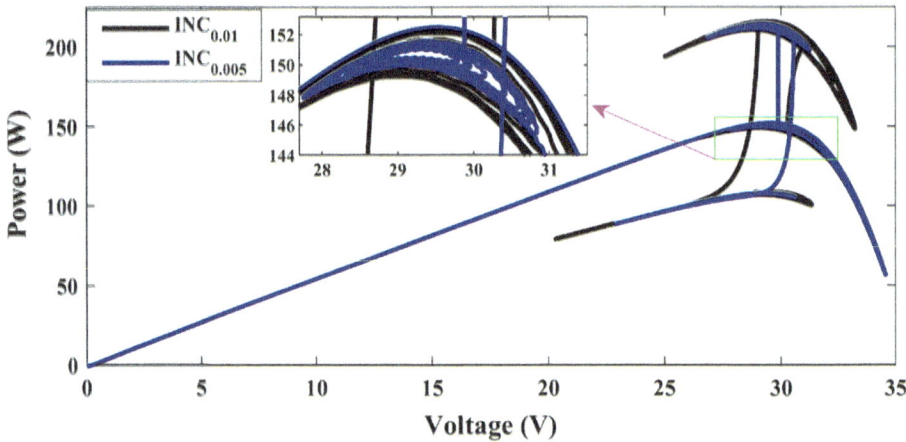

Fig. 3.61 Power versus voltage tracking simulation curves using the INC MPPT method

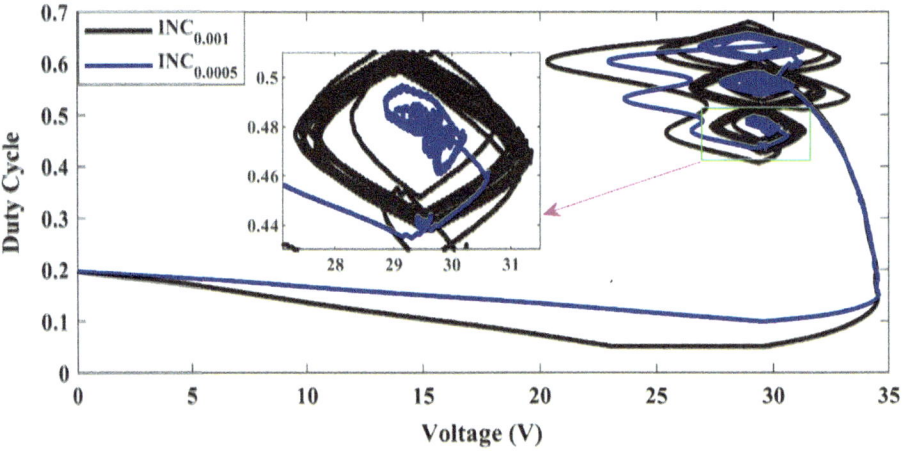

Fig. 3.62 Duty cycle versus voltage tracking simulation curves using the INC MPPT method

3.4 Summary

This chapter has provided an in-depth analysis of the modeling, design, and simulation of a complete standalone solar PV system. The performance of the system was tested using two well-known software tools for photovoltaic system implementation: MATLAB/Simulink and Proteus. The standalone solar PV system was composed of several key components, including a PV module, a DC-DC converter, an MPPT controller, and a DC load.

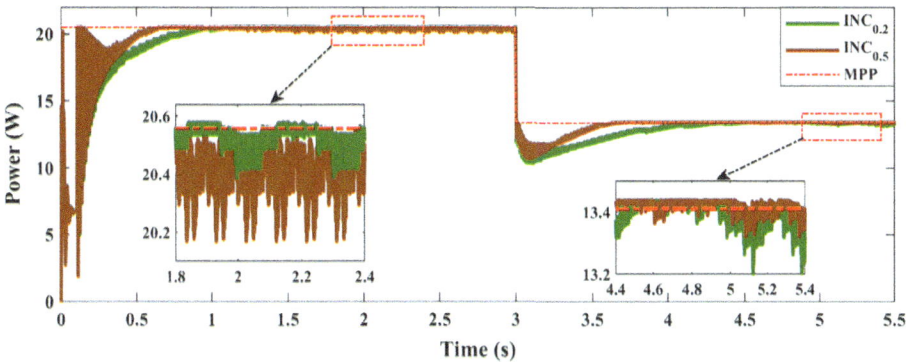

Fig. 3.63 PV output power waveforms obtained using the INC MPPT technique in the Proteus environment

The chapter began with a comprehensive modeling of the PV module in both MATLAB/Simulink and Proteus environments, enabling a better understanding of its operation. Various plots of the P–V and I–V characteristics of the PV module were presented to demonstrate the impact of meteorological factors (such as solar insolation, ambient temperature, and partial shading) as well as electrical parameters (such as ideality factor, saturation current, and shunt/series resistances).

Next, a detailed modeling of various DC-DC converters, including boost, buck, interleaved boost, and three-level boost converters, was conducted using the Simulink environment in MATLAB. This modeling process was essential for the design and selection of appropriate components for the DC-DC converter.

Finally, after modeling the main components of the standalone PV system, simulations were performed to evaluate the complete system's performance. Conventional MPPT techniques, such as P&O and INC methods, were implemented to assess their effectiveness in tracking the maximum power point of the system.

References

1. S. Vighetti, Systèmes photovoltaïques raccordés au réseau: Choix et dimensionnement des étages de conversion (Institut National Polytechnique de Grenoble-INPG, 2010)
2. T. Zhang, H. Yang, High efficiency plants and building integrated renewable energy systems, in *Handbook Energy Efficiency Building: A Life Cycle Approach* (2019), pp. 441–595. https://doi.org/10.1016/B978-0-12-812817-6.00040-1
3. S.M. Islam, C.V. Nayar, A. Abu-Siada, M.M. Hasan, Power electronics for renewable energy sources, in *Power Electronics Handbook* (2018), pp. 783–827. https://doi.org/10.1016/B978-0-12-811407-0.00027-1

4. H. Rezk, A.H.M. El-Sayed, Sizing of a stand alone concentrated photovoltaic system in Egyptian site. Int. J. Electr. Power Energy Syst. **45**(1) (2013). https://doi.org/10.1016/j.ijepes.2012.09.001
5. T.D. Roy, Simulation and analysis of photovoltaic stand-alone systems (2013)
6. V. Salas, Stand-alone photovoltaic systems. Perform. Photovolt. Syst. Model. Meas. Assess. 251–296 (2017). https://doi.org/10.1016/B978-1-78242-336-2.00009-4
7. V. Khanna, B.K. Das, D. Bisht, Vandana, P.K. Singh, A three diode model for industrial solar cells and estimation of solar cell parameters using PSO algorithm. Renew. Energy **78**, 105–113 (2015). https://doi.org/10.1016/J.RENENE.2014.12.072
8. M.N.I. Sarkar, Effect of various model parameters on solar photovoltaic cell simulation: a SPICE analysis. Renew. Wind. Water, Sol. **3**(1) (2016). https://doi.org/10.1186/s40807-016-0035-3
9. T. Ma, H. Yang, L. Lu, Solar photovoltaic system modeling and performance prediction. Renew. Sustain. Energy Rev. **36** (2014). https://doi.org/10.1016/j.rser.2014.04.057
10. H.L. Tsai, Insolation-oriented model of photovoltaic module using matlab/simulink. Sol. Energy **84**(7), 1318–1326 (2010). https://doi.org/10.1016/J.SOLENER.2010.04.012
11. K. Ishaque, Z. Salam, H. Taheri, Simple, fast and accurate two-diode model for photovoltaic modules. Sol. Energy Mater. Sol. Cells **95**(2) (2011). https://doi.org/10.1016/j.solmat.2010.09.023
12. K. Nishioka, N. Sakitani, Y. Uraoka, T. Fuyuki, Analysis of multicrystalline silicon solar cells by modified 3-diode equivalent circuit model taking leakage current through periphery into consideration. Sol. Energy Mater. Sol. Cells **91**(13) (2007). https://doi.org/10.1016/j.solmat.2007.04.009
13. K. Ishaque, Z. Salam, Syafaruddin, A comprehensive MATLAB Simulink PV system simulator with partial shading capability based on two-diode model. Sol. Energy **85**(9) (2011). https://doi.org/10.1016/j.solener.2011.06.008
14. S. Motahhir, A. El Hammoumi, A. El Ghzizal, Photovoltaic system with quantitative comparative between an improved MPPT and existing INC and P&O methods under fast varying of solar irradiation. Energy Rep. **4** (2018). https://doi.org/10.1016/j.egyr.2018.04.003
15. S. Necaibia, M.S. Kelaiaia, H. Labar, A. Necaibia, E.D. Castronuovo, Enhanced auto-scaling incremental conductance MPPT method, implemented on low-cost microcontroller and SEPIC converter. Sol. Energy **180** (2019). https://doi.org/10.1016/j.solener.2019.01.028
16. F.A.O. Aashoor, Maximum power point tracking techniques for photovoltaic water pumping system (University of Bath, 2015)
17. C. Abdelkhalek, E.B. Said, A. Younes, A novel MPPT tactic for PV systems with fast-converging speed and zero oscillation, in *2020 5th International Conference on Renewable Energies for Developing Countries, REDEC 2020* (2020). https://doi.org/10.1109/REDEC49234.2020.9163606
18. H.B. Vika, Modelling of photovoltaic modules with battery energy storage in simulink/matlab: with in-situ measurement comparisons (2014)
19. M. Mahrous Abdelsattar Mahrous, Modelica models of PV elements for system level simulation and control studies (2015)
20. MathWorks, PV Array Block. https://www.mathworks.com/help/physmod/sps/referencelist.html?type=block&category=renewable-energy-relib&s_tid=CRUX_topnav. Accessed 15 Nov 2021
21. M.B. Danoune, A. Djafour, A. Gougui, N. Khelfaoui, H. Boutelli, Characterization of photovoltaic panel using single diode and double diode models a comparative study with experimental validation, in *The 5th International Seminar on New and Renewable Energies, Gharda{\"i}a—Algeria* (2018), pp. 24–25

22. A. Chellakhi, S. El Beid, Y. Abouelmahjoub, Implementation of a novel MPPT tactic for PV system applications on MATLAB/simulink and proteus-based arduino board environments. Int. J. Photoenergy **2021** (2021). https://doi.org/10.1155/2021/6657627
23. C. Abdelkhalek, E.B. Said, A. Younes, Simulation Et Implémentation D'un Système PV sous l'Environnement Proteus Avec Nouvelle Commande MPPT. OSF Preprints (2021)
24. R. Bradai et al., Experimental assessment of new fast MPPT algorithm for PV systems under non-uniform irradiance conditions. Appl. Energy **199** (2017). https://doi.org/10.1016/j.apenergy.2017.05.045
25. E.E. van Dyk, E.L. Meyer, Analysis of the effect of parasitic resistances on the performance of photovoltaic modules. Renew. Energy **29**(3) (2004). https://doi.org/10.1016/S0960-1481(03)00250-7
26. E. Cuce, P.M. Cuce, I.H. Karakas, T. Bali, An accurate model for photovoltaic (PV) modules to determine electrical characteristics and thermodynamic performance parameters. Energy Convers. Manag. **146**, 205–216 (2017). https://doi.org/10.1016/J.ENCONMAN.2017.05.022
27. E. Cuce, P.M. Cuce, T. Bali, An experimental analysis of illumination intensity and temperature dependency of photovoltaic cell parameters. Appl. Energy **111** (2013). https://doi.org/10.1016/j.apenergy.2013.05.025
28. P. Singh, S. N. Singh, M. Lal, M. Husain, Temperature dependence of I–V characteristics and performance parameters of silicon solar cell. Sol. Energy Mater. Sol. Cells **92**(12) (2008). https://doi.org/10.1016/j.solmat.2008.07.010
29. T. Salmi, M. Bouzguenda, A. Gastli, A. Masmoudi, MATLAB/simulink based modelling of solar photovoltaic cell. Int. J. Renew. Energy Res. **2**(2) (2012). https://doi.org/10.20508/ijrer.42248
30. M.F. Nayan, S.M.S. Ullah, Modelling of solar cell characteristics considering the effect of electrical and environmental parameters, in *International Conference on Green Energy and Technology, ICGET 2015* (2015). https://doi.org/10.1109/ICGET.2015.7315096
31. V. Viswanatha, A complete mathematical modeling, simulation and computational implementation of boost converter via MATLAB/Simulink (2017)
32. M.G. Villalva, T.G. De Siqueira, E. Ruppert, Voltage regulation of photovoltaic arrays: small-signal analysis and control design. IET Power Electron. **3**(6) (2010). https://doi.org/10.1049/iet-pel.2008.0344
33. A. Chellakhi, S. El Beid, Y. Abouelmahjoub, An improved adaptable step-size P&O MPPT approach for standalone photovoltaic systems with battery station. Simul. Model. Pract. Theory **121**, 102655 (2022). https://doi.org/10.1016/J.SIMPAT.2022.102655
34. A. Luque, S. Hegedus, *Handbook of Photovoltaic Science and Engineering* (2011). https://doi.org/10.1002/9780470974704
35. V.G.R. Kummara et al., A comprehensive review of DC–DC converter topologies and modulation strategies with recent advances in solar photovoltaic systems. Electronics (Switzerland) **9**(1) (2020). https://doi.org/10.3390/electronics9010031
36. S. Chtita, A. Derouich, A. El Ghzizal, S. Motahhir, An improved control strategy for charging solar batteries in off-grid photovoltaic systems. Sol. Energy **220** (2021). https://doi.org/10.1016/j.solener.2021.04.003
37. C. Abdelkhalek, E.L.B. Said, A. Younes, A. Hassan, A study and implementation of interleaved boost converter with a novel MPPT tactic for PV systems, in *2020 IEEE 2nd International Conference on Electronics, Control, Optimization and Computer Science (ICECOCS)* (2020), pp. 1–6.
38. X. Ruan, B. Li, Q. Chen, S.C. Tan, C.K. Tse, Fundamental considerations of three-level DC-DC converters: topologies, analyses, and control. IEEE Trans. Circuits Syst. I Regul. Pap. **55**(11) (2008). https://doi.org/10.1109/TCSI.2008.927218

39. M. Samadi, S.M. Rakhtala, Reducing cost and size in photovoltaic systems using three-level boost converter based on fuzzy logic controller. Iran. J. Sci. Technol. Trans. Electr. Eng. **43** (2019). https://doi.org/10.1007/s40998-018-0145-6
40. A. Chellakhi, S. El Beid, Y. Abouelmahjoub, Y. Mchaouar, An efficient implementation of three-level boost converter with capacitor voltage balancing for an advanced MPPT approach in PV Systems. e-Prime Adv. Electr. Eng. Electron. Energy **9**(April), 100688 (2024). https://doi.org/10.1016/j.prime.2024.100688
41. L. Liu, X. Meng, C. Liu, A review of maximum power point tracking methods of PV power system at uniform and partial shading. Renew. Sustain. Energy Rev. **53** (2016). https://doi.org/10.1016/j.rser.2015.09.065
42. M.F.N. Tajuddin, M.S. Arif, S.M. Ayob, Z. Salam, Perturbative methods for maximum power point tracking (MPPT) of photovoltaic (PV) systems: a review. Int. J. Energy Res. **39**(9) (2015). https://doi.org/10.1002/er.3289
43. S. Motahhir, A. Chalh, A. El Ghzizal, S. Sebti, A. Derouich, Modeling of photovoltaic panel by using proteus. J. Eng. Sci. Technol. Rev. **10**(2) (2017). https://doi.org/10.25103/jestr.102.02
44. K.S. Tey, S. Mekhilef, Modified incremental conductance MPPT algorithm to mitigate inaccurate responses under fast-changing solar irradiation level. Sol. Energy **101** (2014). https://doi.org/10.1016/j.solener.2014.01.003
45. A. Chellakhi, S. El Beid, M. El Marghichi, E.M. Bouabdalli, A. Harrison, H. Abouobaida, Implementation of a low-cost current perturbation-based improved PO MPPT approach using Arduino board for photovoltaic systems. e-Prime Adv. Electr. Eng. Electron. Energy **10**(September), 100807 (2024). https://doi.org/10.1016/j.prime.2024.100807
46. M. Gopahanal Manjunath, C. Vyjayanthi, C.N. Modi, Adaptive step size based drift-free P&O algorithm with power optimiser and load protection for maximum power extraction from PV panels in stand-alone applications. IET Renew. Power Gener. **15**(6) (2021). https://doi.org/10.1049/rpg2.12105
47. N.S. D'Souza, L.A.C. Lopes, X.J. Liu, Comparative study of variable size perturbation and observation maximum power point trackers for PV systems. Electr. Power Syst. Res. **80**(3) (2010). https://doi.org/10.1016/j.epsr.2009.09.012
48. M.A. Elgendy, B. Zahawi, D.J. Atkinson, Operating characteristics of the P&O algorithm at high perturbation frequencies for standalone PV systems. IEEE Trans. Energy Convers. **30**(1) (2015). https://doi.org/10.1109/TEC.2014.2331391
49. A. Loukriz, M. Haddadi, S. Messalti, Simulation and experimental design of a new advanced variable step size incremental conductance MPPT algorithm for PV systems. ISA Trans. **62** (2016). https://doi.org/10.1016/j.isatra.2015.08.006
50. Q. Mei, M. Shan, L. Liu, J.M. Guerrero, A novel improved variable step-size incremental-resistance MPPT method for PV systems. IEEE Trans. Ind. Electron. **58**(6) (2011). https://doi.org/10.1109/TIE.2010.2064275
51. S. Motahhir, Contribution to the optimization of energy withdrawn from a PV panel using an embedded system. SSRN Electron. J. (2021). https://doi.org/10.2139/ssrn.3804081

4. Principle and Simulation Investigation of The Newly Proposed MPPT Approaches

4.1 Introduction

Enhancing the efficiency of PV systems, particularly standalone systems, requires maximizing the power output from the PV panel or array. Among the various techniques employed for optimizing PV systems, one of the most widely recognized is MPPT. This technique ensures that the system operates at maximum power, achieved by maintaining optimal voltage and current levels. This is facilitated by a DC-DC converter, which serves as an impedance adapter between the PV generator and the load. The converter continuously monitors the PV panel/array's power output and adjusts the pulse-width modulation (PWM) signal to regulate the duty cycle, as illustrated in Fig. 4.1.

As discussed in Chap. 3, numerous MPPT strategies have been explored in the literature, with the most prevalent being INC and P&O methods. MPPT strategies can be categorized into three main groups:

(i) Conventional methods such as P&O, INC, Fractional Open-Circuit Voltage (FOCV), and Fixed Step Current Control (FSCC);
(ii) Soft computing methods including FLC, ANN, PSO, and GA;
(iii) Hybrid methods, which combine different techniques, such as PSO-P&O, FLC-INC, ANN-Hill Climbing (HC), and Fuzzy PSO (FPSO) [1].

Each of these methods offers specific advantages and drawbacks. However, several critical challenges persist, such as the *drift effect* during rapid changes in insolation, temperature fluctuations, and load variations, as well as instability around the MPP during steady-state operation, as depicted in Fig. 4.2 [2, 3]. To address these challenges, this chapter introduces and analyzes four novel MPPT approaches. These approaches are designed to

Fig. 4.1 A schematic overview of the block diagram for a standalone PV system equipped with an MPPT controller

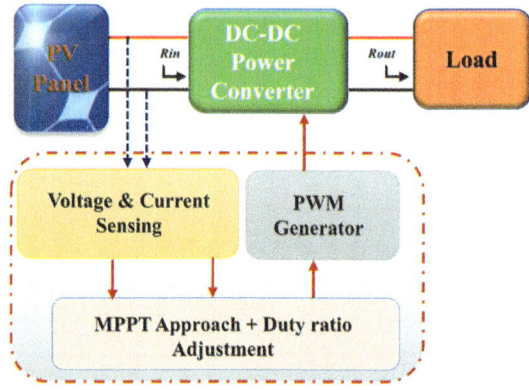

Fig. 4.2 Schematic representation of the key challenges faced by conventional MPPT methods

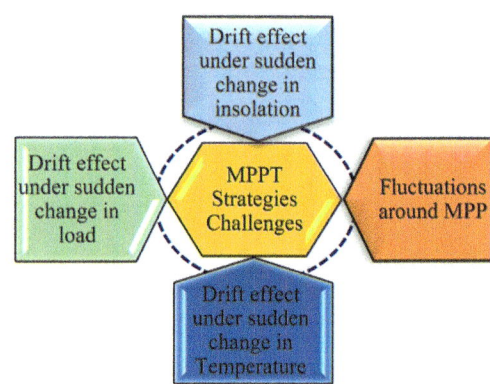

meet key criteria such as simplicity, low-cost implementation, fast dynamic response, and high tracking efficiency, as discussed in Chap. 3.

- The first approach addresses MPP tracking challenges during rapid temperature fluctuations.
- The second approach is designed to handle sudden changes in solar irradiance and load conditions.
- The third approach aims to enhance tracking performance across a range of climatic conditions.
- The fourth and final approach is intended to cope with diverse temperature variability.

This chapter explores the theoretical principles of the proposed MPPT methods and their simulation validation.

4.2 First Proposed Algorithm: An Enhanced MPPT Approach for Temperature Variation

As discussed in Chap. 3, temperature variations can significantly impact the performance of a PV module, particularly its maximum available power. This relationship is clearly illustrated in the PV module's P-V characteristics in Fig. 4.3, which shows a substantial decrease in MPP as the temperature increases under various levels of insolation.

To address this issue, the first enhanced MPPT approach is designed to improve the tracking accuracy of conventional MPPT methods under rapid temperature changes. This new MPPT algorithm provides fast tracking speed, reduced steady-state oscillation, and can be integrated with existing MPPT strategies to improve their precision, convergence time, and minimize power losses. Despite these improvements, the enhanced MPPT method maintains a relatively simple implementation without added complexity, making it both fast and accurate in finding the MPP under severe temperature fluctuations.

In standalone PV systems, DC-DC converters are critical to the functioning of MPPT algorithms. Without these converters, MPPT methods cannot be applied, and the PV module will not reach its MPP. Among the types of DC-DC converters, the boost converter (shown in Fig. 3.32) is most commonly used in MPPT controller applications [4–7].

The voltage and current relationships on the input and output sides of the boost converter are governed by the following Eqs. (4.1)–(4.3):

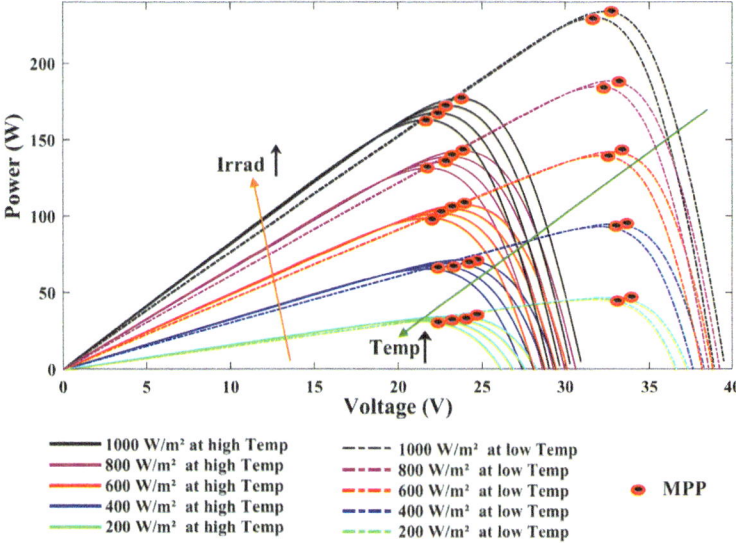

Fig. 4.3 Illustration of the changes in the MPP on the P-V curves in response to temperature variations at different levels of solar radiation

$$V_{pv} = \frac{V_{out}}{C(d)} \tag{4.1}$$

$$I_{pv} = I_{out} * C(d) \tag{4.2}$$

where $C(d)$ is defined as:

$$C(d) = \frac{1}{1-d} \tag{4.3}$$

By dividing Eqs. (4.1) by (4.2), we derive:

$$R_{in} = \frac{V_{pv}}{I_{pv}} = \frac{1}{C(d)^2} \times \frac{V_{out}}{I_{out}} = \frac{R_{out}}{C(d)^2} \tag{4.4}$$

which simplifies to:

$$R_{pv} = R_{in} = \frac{R_{out}}{C(d)^2} = \frac{R_{load}}{C(d)^2} \tag{4.5}$$

where R_{pv} denotes the resistance seen by the PV string and R_{load} denotes the load resistance, calculate as:

$$R_{load} = \frac{1}{(1-d)^2} \times \frac{V_{pv}}{I_{pv}} \tag{4.6}$$

At a specific instant (k), Eq. (4.6) becomes:

$$R_{load}(k) = \frac{1}{(1-d(k))^2} \times \frac{V_{pv}(k)}{I_{pv}(k)} \tag{4.7}$$

Thus, the boost converter's duty ratio can be derived as:

$$d(k+1) = 1 - \frac{\sqrt{V_{pv}(k)/I_{pv}(k)}}{\sqrt{R_{load}(k)}} \tag{4.8}$$

4.2.1 Principle

The principle behind this approach is based on the current-voltage (I-V) characteristics of the PV module under rapid temperature changes, as illustrated in Figs. 4.4 and 4.5. When the temperature rises from a lower value (25 °C) to a higher value (50 °C), the operating point (OP) of the PV module shifts from point A (V_{mpp1}, I_{mpp1}), which is the MPP at 25 °C, to point B ($V(k), I(k)$) along load line 1. At this stage, the OP is quite distant from point E (V_{mpp2}, I_{mpp2}), the expected MPP at 50 °C [8, 9]. The improved

4.2 First Proposed Algorithm: An Enhanced MPPT Approach ...

MPPT approach adjusts the OP to point C $(V(k), I(k-1))$ on load line 2, calculated using Eq. (4.7) (discussed above). By substituting the actual voltage $V(k)$, the previous current $I(k-1)$ (approximately the MPP current at 25 °C), and $R_{out}(k)$ into Eq. (4.8), the OP is shifted to point D on load line 2, as shown in Fig. 4.6. A conventional MPPT technique such as P&O, INC, or HC is then employed to track the new MPP at 50 °C in a few steps.

In the case of a rapid temperature decrease, the same process is applied in reverse. As shown in Figs. 4.5 and 4.7, when the OP shifts from point E (V_{mpp2}, I_{mpp2}) to point D $(V(k), I(k))$ along load line 3, the proposed approach moves the OP to point C

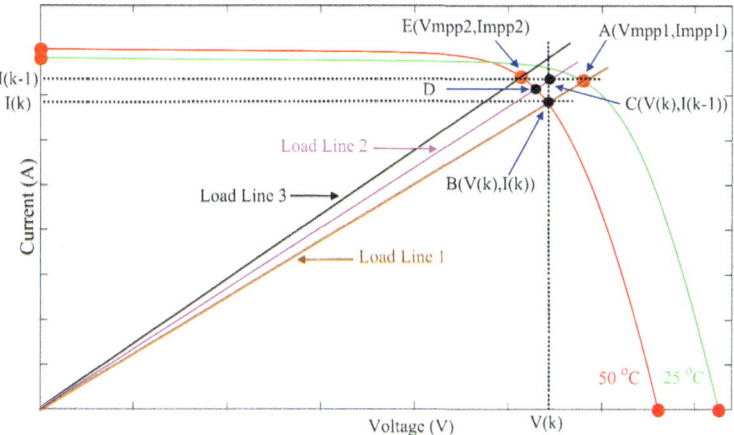

Fig. 4.4 The movement of the OP during a rapid increase in temperature

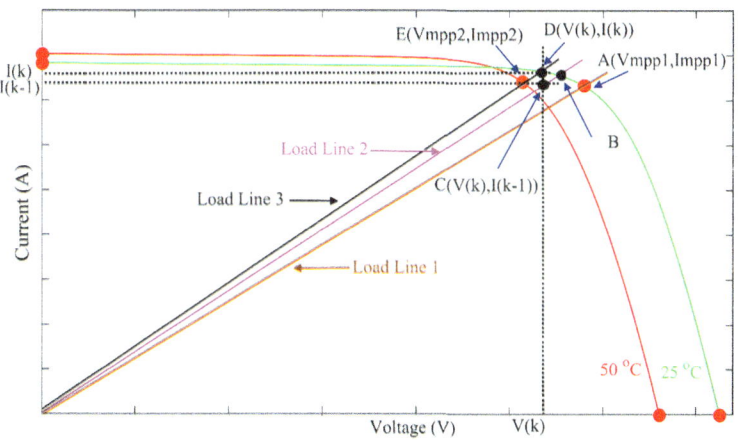

Fig. 4.5 The movement of the OP during a rapid decrease in temperature

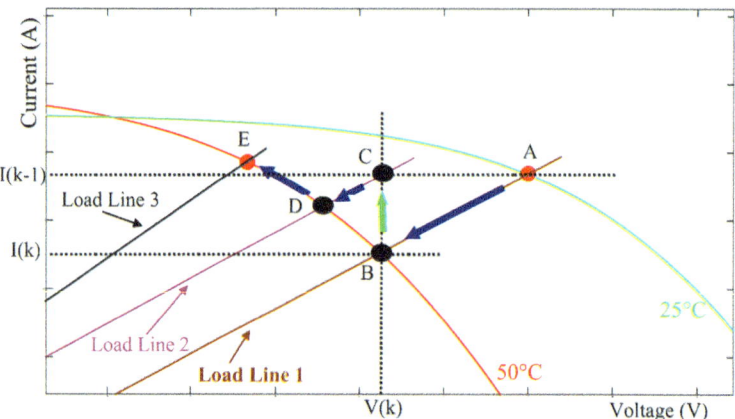

Fig. 4.6 A detailed view of the OP movement procedure during a rapid increase in temperature

$(V(k), I(k-1)(I_{mpp2}))$, calculated using Eq. (4.7). A new duty ratio is then determined using Eq. (4.8) by substituting the current values, leading the OP to operate near the MPP at 25 °C (point B). Finally, a conventional MPPT method is used to track the new MPP in a few steps [8, 9].

Figure 4.8 shows the flowchart of the enhanced MPPT approach. Initially, the temperature condition is detected using Eq. (4.9). If no significant temperature change is observed, the MPP is tracked using a conventional MPPT technique such as INC or P&O. However, if there is a temperature variation, Eqs. (4.7) and (4.8) are used to closely follow the new MPP with a high convergence rate.

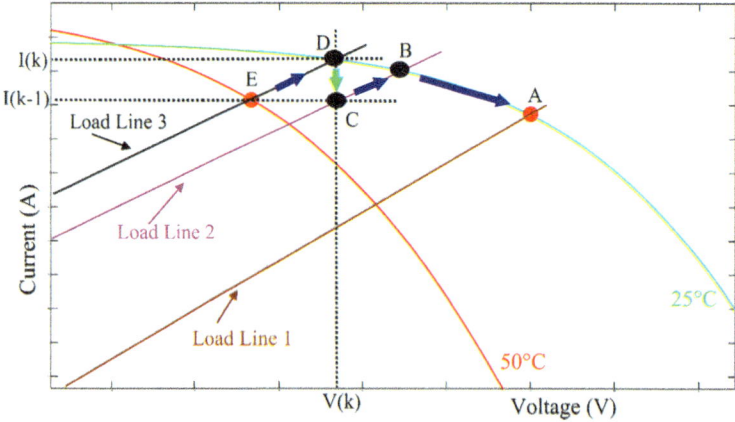

Fig. 4.7 A detailed view of the OP movement procedure during a rapid decrease in temperature

4.2 First Proposed Algorithm: An Enhanced MPPT Approach …

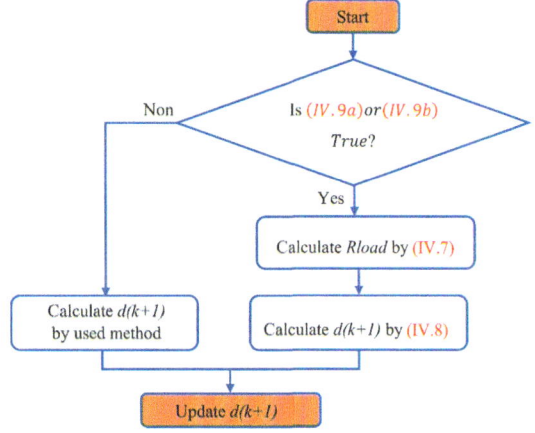

Fig. 4.8 Flowchart of the proposed enhanced MPPT approach

$$\left.\begin{array}{l}\Delta P \text{ and } \Delta V < 0, \Delta I < 0 \text{ Temperature increases } (a)\\ \Delta P \text{ and } \Delta V > 0, \Delta I < 0 \text{ Temperature decreases } (b)\end{array}\right\} \quad (4.9)$$

4.2.2 MATLAB/Simulink Implementation and Simulation Results

To evaluate and compare the performance of the improved MPPT approach with other methods such as P&O, INC, and the Modified MPP-Locus technique from [10], a series of simulations were performed using the Simulink environment in MATLAB. The general standalone PV system used for these simulations is depicted in Fig. 4.9.

The temperature profile applied in the simulations is shown in Fig. 4.10, where two irradiance levels—1 and 0.6 kW/m² —were considered. The parameters of the complete PV system, including the PV module (1Soltech 1STH-215-P) and the DC-DC boost converter, are listed in Table 4.1.

4.2.2.1 Performance of the Improved MPPT Approach Compared to the P&O MPPT Scheme

Figure 4.11 shows the simulation results of power, voltage, current, and duty cycle for the improved and traditional P&O approaches under temperature variations and two levels of irradiance. Figure 4.12 provides an analysis of the steady-state error in power and voltage under Standard Test Conditions (STC). Meanwhile, Figs. 4.13 and 4.14 compare the response time and tracking efficiency of the improved MPPT approach with the traditional P&O scheme.

Based on the simulation results, it can be concluded that the improved MPPT approach significantly outperforms the traditional P&O method. Specifically, it exhibits better performance in terms of:

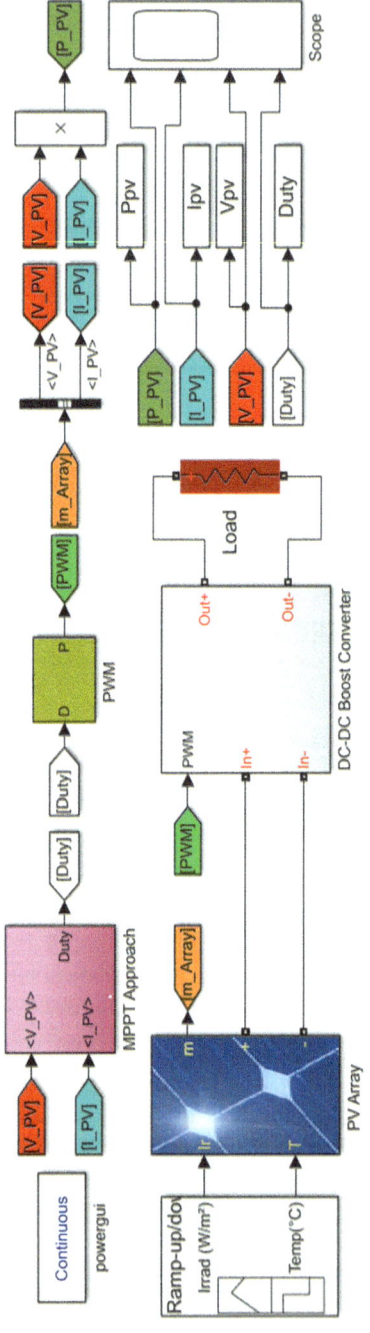

Fig. 4.9 Circuit of the proposed standalone PV system with the suggested MPPT approach

4.2 First Proposed Algorithm: An Enhanced MPPT Approach ...

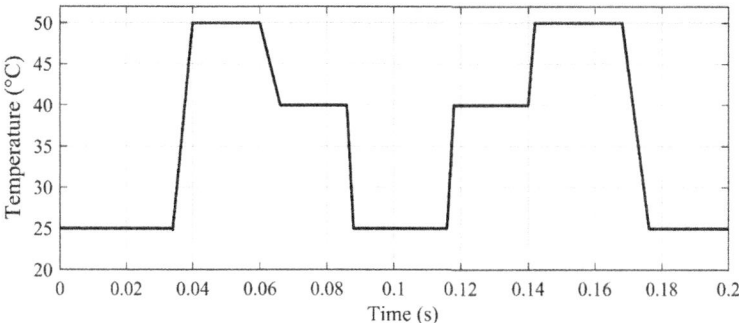

Fig. 4.10 Temperature variation profile

Table 4.1 Parameters of the complete standlone PV system (PV module (1Soltech 1STH-215-P) in the STC and DC-DC boost converter)

Parameters	Variable	Value
Photovoltaic panel		
Power at MPP	P_{MPP}	213 W
MPP voltage	V_{MPP}	29 V
MPP current	I_{MPP}	7.35 A
Voltage of open circuit	V_{oc}	36.3 V
Current of short circuit	I_{sc}	7.84 A
DC-DC boost converter		
Input capacitor	C_{in}	47 µF
Output capacitor	C_{out}	470 µF
Inductor	L	1 mH
Frequency	f	10 kHz
Resistive Load	R_{load}	30 Ω

- **Reduced fluctuations around the MPP**: The improved approach shows far less oscillation around the optimal power, voltage, and current values, as seen in Figs. 4.11 and 4.12.
- **Faster response time**: The improved method responds more quickly to changes in temperature, as indicated by the comparison in Fig. 4.13.
- **Minimized power losses**: The improved MPPT approach achieves lower power losses, especially under varying temperature conditions.
- **Enhanced tracking efficiency**: As shown in Fig. 4.14, the improved method offers significantly higher tracking efficiency compared to the traditional P&O approach.

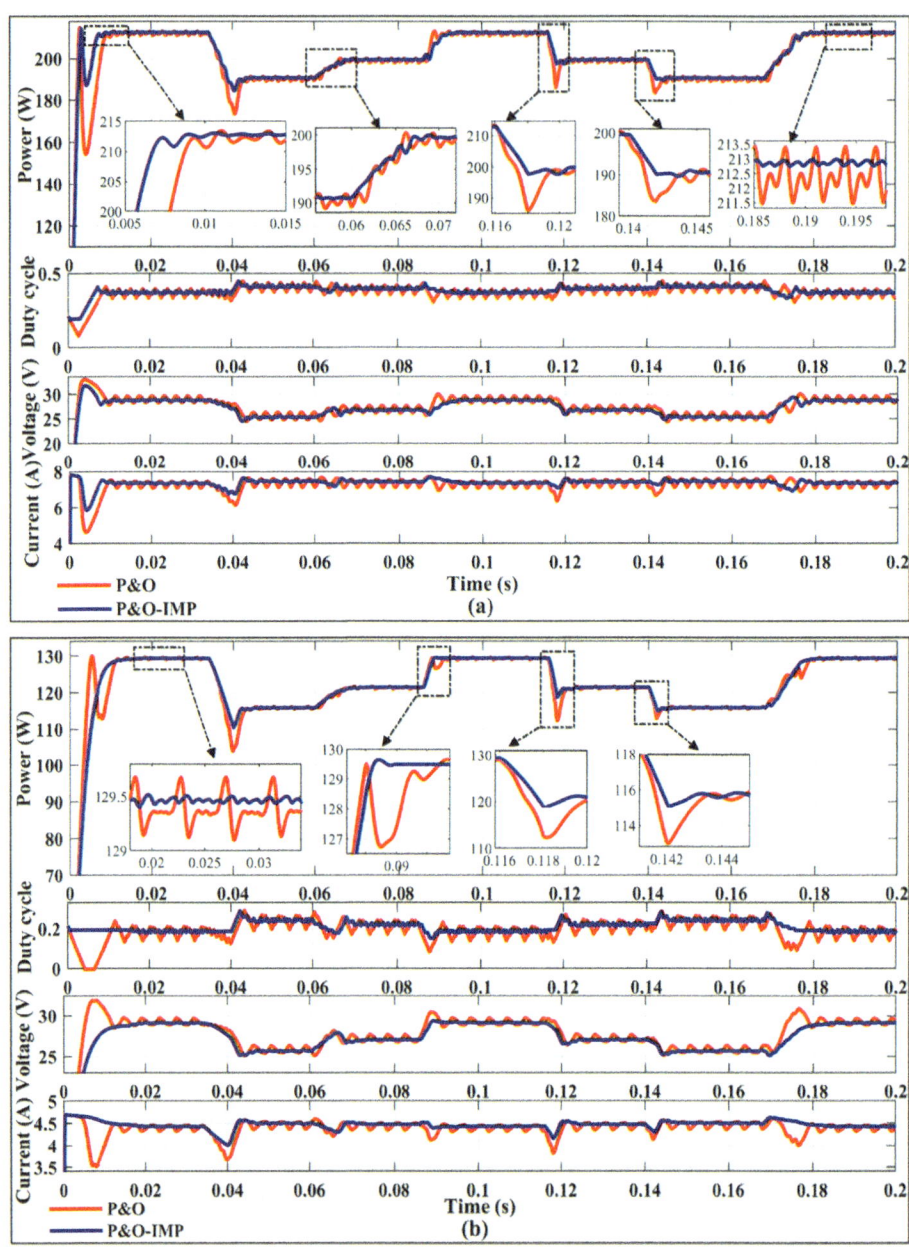

Fig. 4.11 Simulation results of the improved MPPT approach (P&O-IMP) compared to the traditional P&O MPPT algorithm under temperature variation for fixed irradiance levels: **a** 1 kW/m^2 and **b** 0.6 kW/m^2

4.2 First Proposed Algorithm: An Enhanced MPPT Approach ...

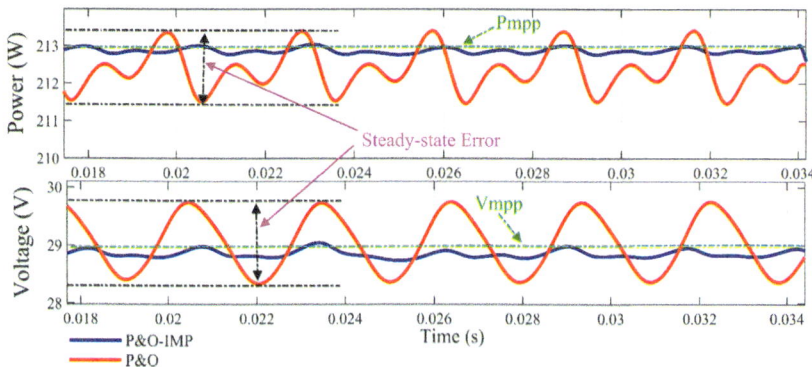

Fig. 4.12 Steady-state error analysis of power and voltage under STC

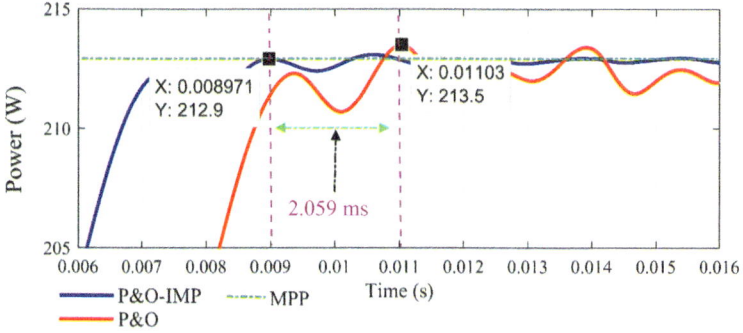

Fig. 4.13 Comparison of response time under STC

In summary, the improved MPPT approach demonstrates superior performance in reducing power losses, minimizing fluctuations, and improving tracking efficiency, particularly under conditions of temperature variation.

4.2.2.2 Performance of the Improved MPPT Approach Compared to the INC MPPT Scheme

The simulation results comparing the traditional Incremental Conductance (INC) MPPT technique with the improved INC approach (INC-IMP) are presented in Fig. 4.15. These results include waveforms for power, voltage, current, and duty cycle under a temperature variation scenario at two fixed irradiance levels: 1 and 0.6 kW/m^2.

Additionally, Fig. 4.16 shows the tracking efficiency waveforms for both the improved and conventional INC MPPT techniques. Based on these results, the improved INC approach (INC-IMP) demonstrates superior performance:

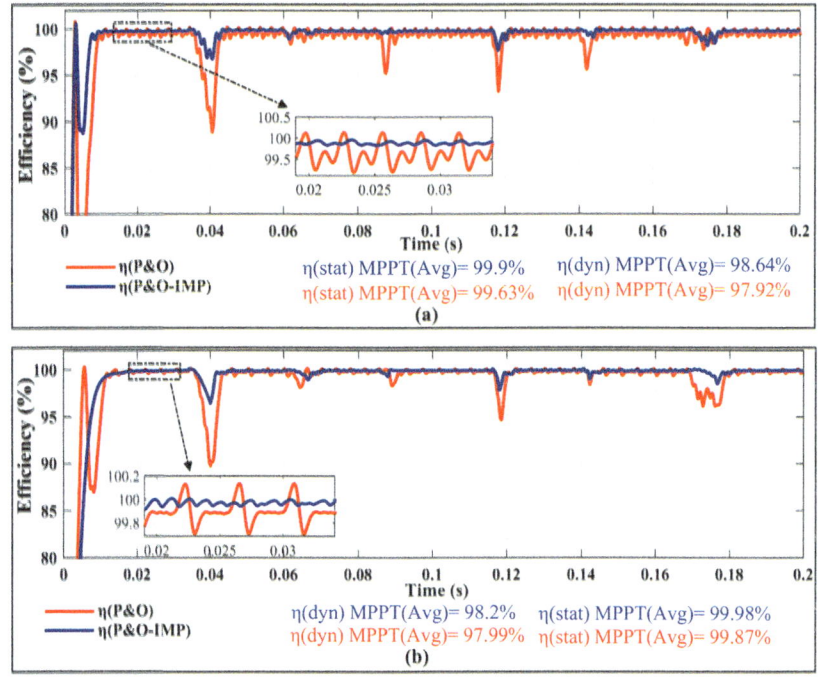

Fig. 4.14 Results of tracking efficiency for the improved MPPT approach (P&O-IMP) compared to the traditional P&O algorithm under temperature variation for fixed irradiance levels: **a** 1 kW/m^2 and **b** 0.6 kW/m^2

- **Higher tracking accuracy**: The improved INC-IMP method tracks the maximum power point (MPP) with greater precision, especially under varying temperature conditions.
- **Reduced power losses**: The improved method minimizes power losses, particularly during temperature fluctuations, compared to the conventional INC technique.
- **Enhanced static and dynamic tracking efficiency**: As shown in Fig. 4.16, the improved INC-IMP significantly improves both static and dynamic tracking efficiency, ensuring more stable and efficient operation.

4.2.2.3 Performance of the Improved MPPT Approach Compared to the Modified MPP-Locus MPPT Scheme

The comparison between the improved MPPT approach (Mod-MPP-Locus-IMP) and the traditional Modified MPP-Locus MPPT technique is shown in Figs. 4.17 and 4.18. These simulations were conducted under temperature variation with two fixed irradiance levels: 1 and 0.6 kW/m^2.

Figure 4.17 presents waveforms for power, voltage, current, and duty cycle, highlighting the differences between the two approaches. The improved Mod-MPP-Locus-IMP

4.2 First Proposed Algorithm: An Enhanced MPPT Approach ...

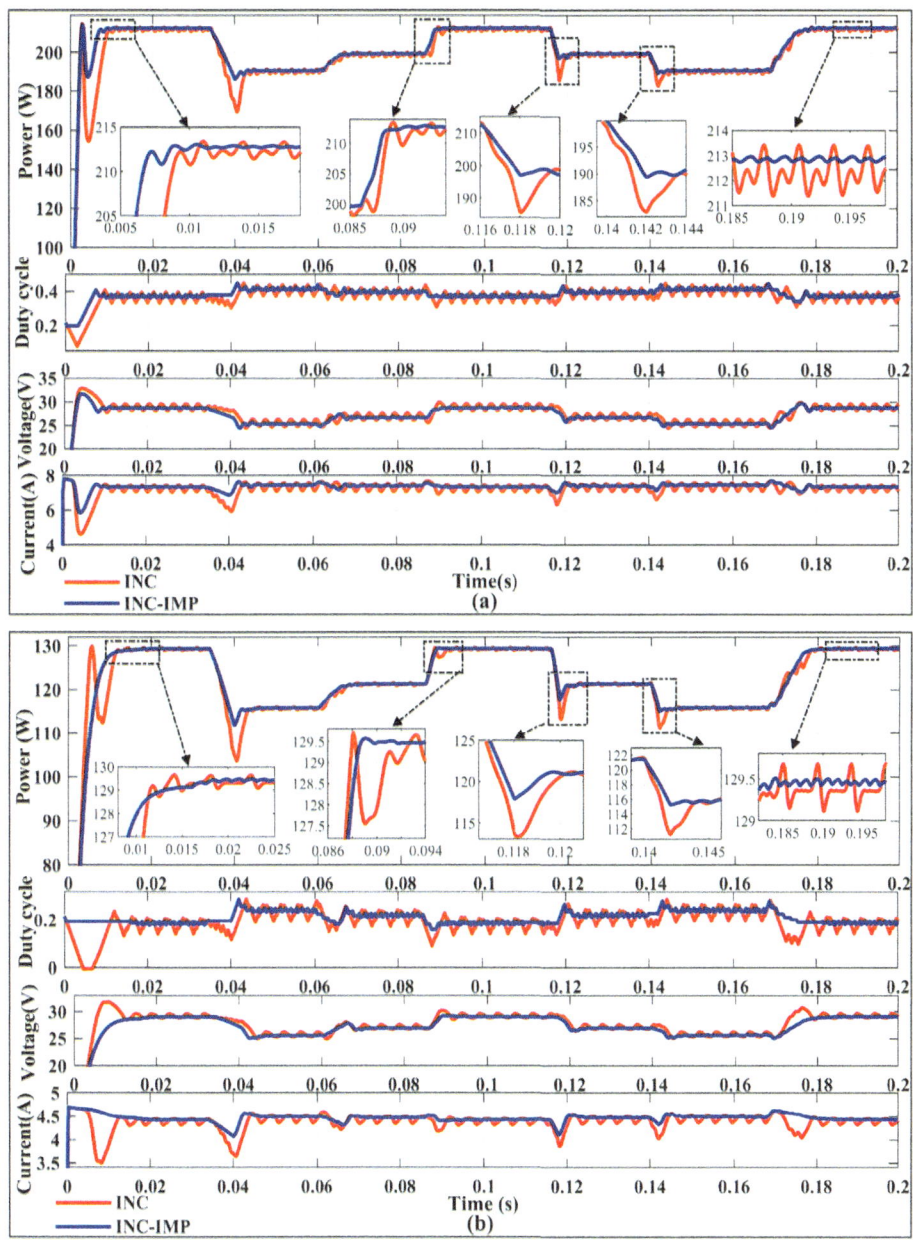

Fig. 4.15 Simulation results of the improved MPPT approach (INC-IMP) compared to those of the INC MPPT technique with regard to the temperature variability under fixed irradiation: **a** 1 kW/m^2 and **b** 0.6 kW/m^2

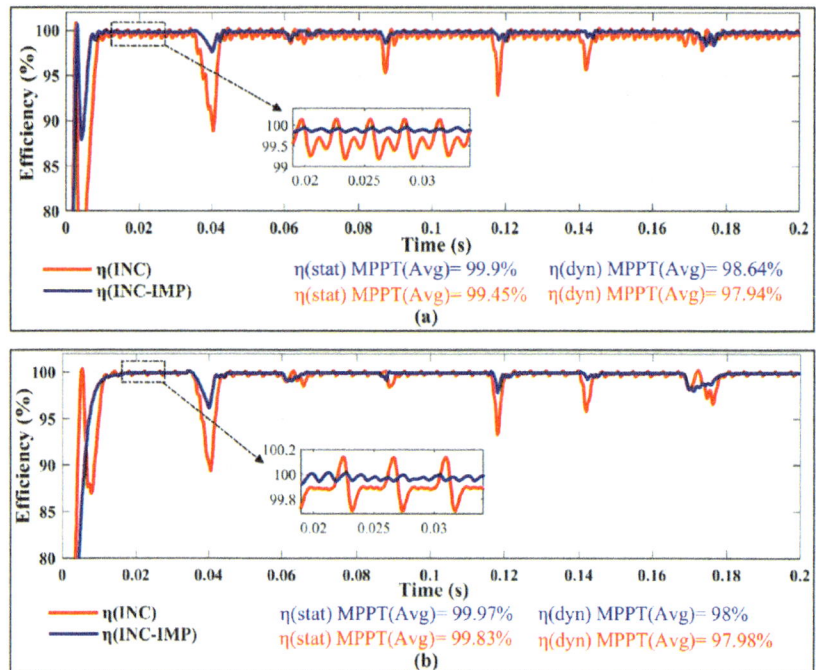

Fig. 4.16 Results of the tracking efficiency of the improved MPPT approach (INC-IMP) compared to the INC MPPT technique regarding the variability of temperature under fixed irradiation: **a** 1 kW/m^2 and **b** 0.6 kW/m^2

approach exhibits fewer fluctuations during the temperature variation scenario and reaches the MPP more accurately, even under rapid temperature changes.

Figure 4.18 shows the tracking efficiency curves. The improved MPPT method demonstrates higher tracking efficiency under both static and dynamic conditions compared to the traditional Modified MPP-Locus technique.

Key observations from the simulation results include:

- **Reduced fluctuations**: The improved Mod-MPP-Locus-IMP approach shows significantly fewer overshoots in the power curve during rapid temperature changes, resulting in lower power losses compared to the conventional Mod-MPP-Locus method (as shown in Fig. 4.17).
- **Increased tracking efficiency**: The improved approach provides higher tracking efficiency across varying temperature conditions, as shown in Fig. 4.18.

4.2 First Proposed Algorithm: An Enhanced MPPT Approach ...

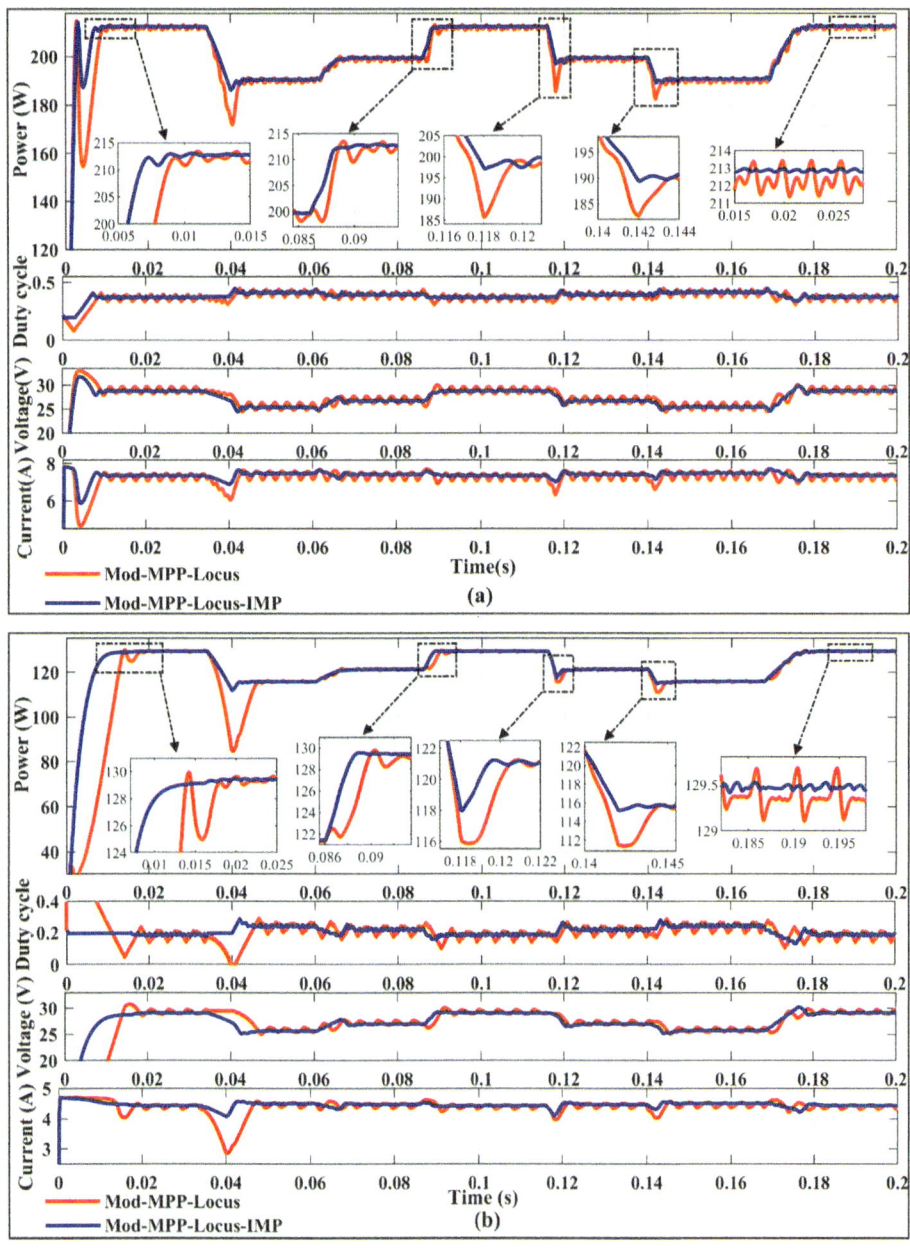

Fig. 4.17 Simulation results of the improved MPPT approach (Mod-MPP-Locus-IMP) compared to the Mod-MPP-Locus MPPT technique regarding temperature varying under fixed irradiation: **a** 1 kW/m^2 and **b** 0.6 kW/m^2

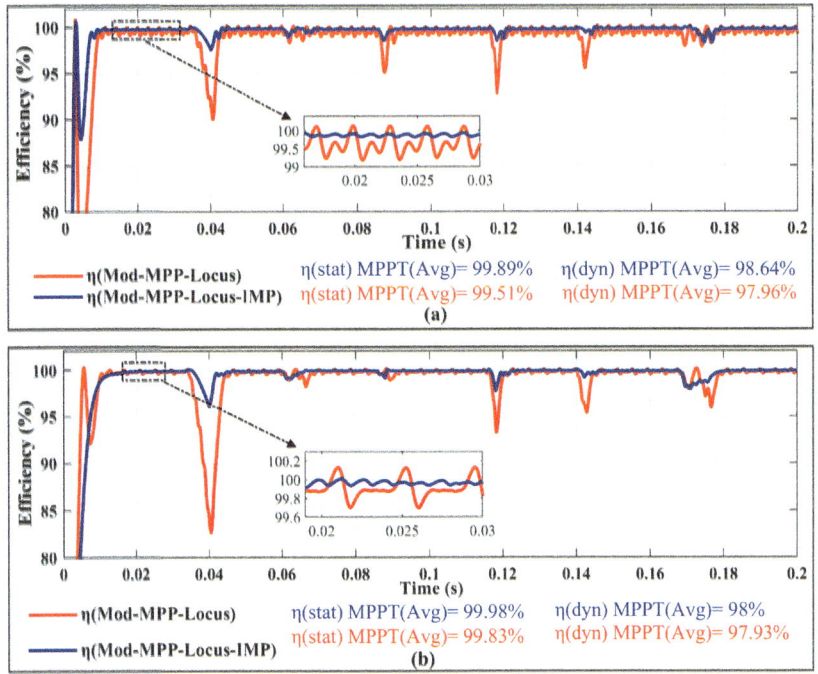

Fig. 4.18 Results of tracking efficiency of the improved MPPT approach (Mod-MPP-Locus-IMP) compared to the Mod-MPP-Locus MPPT technique regarding temperature varying under fixed irradiation: **a** 1 kW/m^2 and **b** 0.6 kW/m^2

A comprehensive comparison of the performance of the improved and conventional MPPT schemes is summarized in Table 4.2. The results demonstrate that the improved MPPT approaches offer significant advantages over conventional methods, particularly under temperature variation scenarios. Key improvements include:

- Higher accuracy in tracking the MPP
- Faster response times
- Lower power losses
- Enhanced efficiency under both static and dynamic conditions

In conclusion, the improved MPPT techniques provide substantial performance gains, making them a more effective solution for optimizing photovoltaic systems in environments with variable temperature conditions.

Table 4.2 Summarize of performances of the improved MPPT approaches compared to the conventional ones

Performance	MPPT techniques					
	P&O	P&O-IMP	INC	INC-IMP	Mod-Locus MPPT	Mod-Locus MPPT-IMP
Tracking speed	Medium	Faster	Medium	Faster	Slow	Faster
Steady-state fluctuation	Large	Small	Medium	Small	Medium	Small
Accuracy/ precision	Low	High	Medium	High	Medium	High
Dynamic efficiency range (%)	97.92–97.99	98.2–98.64	97.94–98	98–98.64	97.93–98	98–98.64
Static efficiency range (%)	99.63–99.87	99.9–99.98	99.45–99.83	99.9–99.7	99.51–99.83	99.89–99.98
Response time (ms)	11.03	8.97	9.77	4.52	19.47	16.9
Power steady-state error (W)	2	Negligible	1	Negligible	0.6	Negligible
Voltage steady-state error (V)	1.5	Negligible	1	Negligible	1	Negligible
Power overshoot	High	Insignificant	High	Insignificant	High	Insignificant

4.3 Second Proposed Algorithm: A Novel MPPT Tactic with Fast Tracking Speed and Zero Oscillation

Numerous studies have demonstrated the development of a wide array of MPPT strategies [1, 11]. Among these, the P&O, INC, and HC methods are particularly prevalent due to their relatively low implementation cost and simplicity. However, these conventional techniques suffer from significant drawbacks, particularly when exposed to sudden changes in meteorological conditions, such as fluctuations in irradiance and ambient temperature. These limitations directly affect their overall efficiency and robustness.

To address these shortcomings, a variety of advanced MPPT techniques, including soft computing methods and hybrid approaches, have been introduced in the scientific literature [1, 11]. While these methods offer potential improvements, their practical applicability is often limited by several constraints, such as high computational requirements, the need for costly hardware, and the use of expensive measurement sensors [12]. These factors restrict their widespread adoption, especially in cost-sensitive applications.

In light of these challenges, a novel MPPT strategy has been developed to enhance the performance of photovoltaic systems under challenging irradiance and load conditions. The proposed method ensures effective power extraction while maintaining low system complexity and providing a cost-effective implementation pathway. Importantly, this approach achieves improved performance without the need for excessive computational resources or costly components. The validity and effectiveness of the proposed approach have been discussed in a recently published journal article [13].

4.3.1 Principle

The proposed MPPT strategy is a simple, direct control method inspired by the P-V characteristics of PV modules, as shown in Fig. 4.19. The MPPs are typically located within a narrow voltage range, constrained by lower and upper MPP voltages V_{Upper} and V_{Lower} under different irradiance levels [13–15].

As illustrated in the flowchart in Fig. 4.20, the proposed method begins by sensing the PV module's current and voltage. It then compares the measured voltage $V(k)$ with the predefined voltage limits V_{Upper} and V_{Lower}. If the measured voltage exceeds V_{Upper}, the voltage is set to V_{Upper}; if it is below V_{Lower}, it is set to V_{Lower}. The next step involves calculating the instantaneous duty cycle $d(k + 1)$ for the boost converter using Eq. (4.8). If $V(k)$ remains within the voltage range, the new duty ratio $d(k + 1)$ is set equal to the previous duty ratio $d(k)$. This approach allows the novel MPPT strategy to rapidly converge to the MPP zone with high accuracy and minimal oscillation [15].

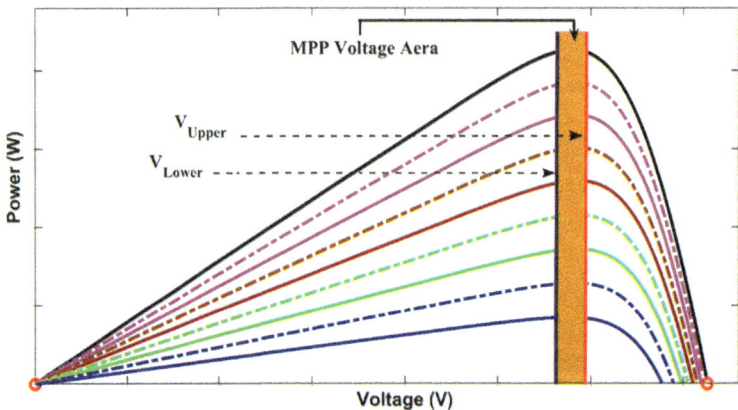

Fig. 4.19 MPP voltage area illustration on P-V curves

4.3 Second Proposed Algorithm: A Novel MPPT Tactic with Fast ...

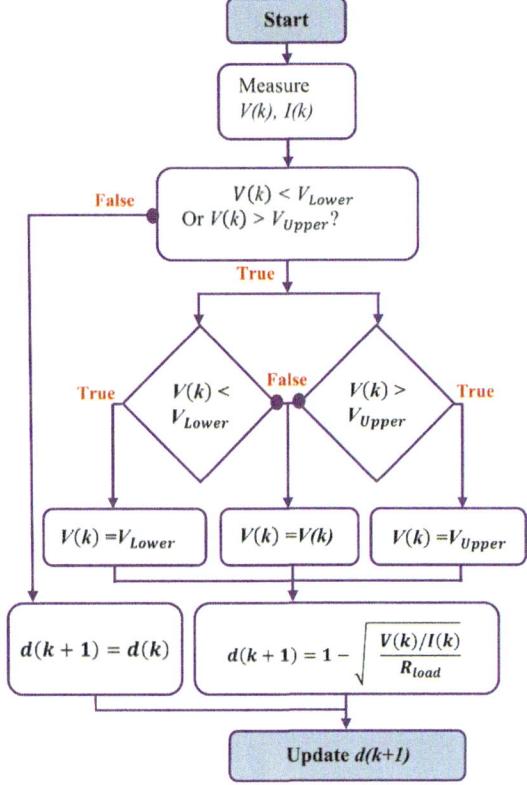

Fig. 4.20 Flowchart of the second proposed MPPT approach [15]

4.3.2 Implementation and Simulation Results in MATLAB/Simulink Environment

The implementation of the newly proposed MPPT strategy for the complete PV system was conducted using two widely recognized tools in PV system engineering simulation: MATLAB/Simulink and Proteus software. MATLAB/Simulink was selected due to its simplicity, flexibility, and ease of use, as well as the availability of a specialized library called SimPowerSystems, which is specifically designed for power systems, including PV systems. Proteus software, on the other hand, was utilized to demonstrate the feasibility of implementing the proposed MPPT strategy using low-cost microcontrollers, such as an Arduino board. Additionally, Proteus allows the use of the same components that would be employed in practical validation, thereby bridging the gap between simulation and real-world implementation.

To evaluate the performance, robustness, and validity of the novel MPPT approach, a comparative analysis was carried out against conventional controllers such as P&O, INC, and FLC MPPT controllers under varying solar irradiance conditions.

Figure 4.21 depicts the implementation of a standalone PV system in MATLAB/Simulink, comprising a PV panel, boost converter, MPPT controller, and load. The specifications for the PV panel and boost converter are consistent with those used in previous simulations involving enhanced MPPT strategies (refer to Table 4.1).

4.3.2.1 Simulation Results in the Presence of Solar Irradiance Varying

To demonstrate the precision and performance of the novel MPPT strategy in tracking the MPP under different insolation conditions, two test scenarios were selected. In the first scenario, the solar irradiance varied suddenly between four levels: 0.8, 1.2, 0.4, and 0.9 kW/m^2, as shown in Fig. 4.22a. In the second scenario, the solar irradiance fluctuated gradually in a sinusoidal pattern between 1.1 and 0.5 kW/m^2, as depicted in Fig. 4.22b. Throughout these tests, the temperature was held constant at 25 °C.

The simulation results for both insolation conditions are presented in Figs. 4.23 and 4.24. The PV power tracked by each MPPT method is illustrated in Fig. 4.23, while Fig. 4.24 presents the waveforms of PV current, PV voltage, and the duty cycle.

Based on the results, it is evident that the novel MPPT strategy outperforms the conventional approaches in tracking the MPP under both sudden and gradual variations in insolation. Specifically, the FLC controller performed the worst, failing to effectively track the MPP in most cases, as depicted in Fig. 4.23. The novel MPPT approach, on the other hand, demonstrated superior performance, precisely tracking the MPP with minimal ripple, thereby reducing power loss and enhancing tracking efficiency. Moreover, the proposed approach exhibited faster convergence and response time, especially under rapid changes in irradiance.

Although the P&O and INC methods were able to track the MPP under both test scenarios, they displayed significant steady-state oscillations, particularly at higher insolation levels. This oscillatory behavior, coupled with their slower response time, led to lower tracking efficiency and increased power loss compared to the proposed approach.

As illustrated in Fig. 4.24, the proposed MPPT strategy effectively adjusted the duty ratio to follow the profile of insolation changes in both test scenarios. This adaptive behavior was further evidenced by the stable PV output current and voltage, which exhibited negligible steady-state ripple.

The tracking performance of the three MPPT techniques was also analyzed through power versus voltage (P-V) and duty versus voltage (D-V) curves under both sudden and sinusoidal insolation changes, as shown in Figs. 4.25, 4.26, 4.27, 4.28, 4.29 and 4.30. The novel MPPT approach demonstrated a well-aligned tracking path close to the MPP of the PV panel in both scenarios, as depicted in Figs. 4.25 and 4.28. In contrast, the P&O and INC methods exhibited a more dispersed MPP trajectory with significant oscillations, as shown in Figs. 4.26, 4.27, 4.29, and 4.30.

4.3 Second Proposed Algorithm: A Novel MPPT Tactic with Fast ...

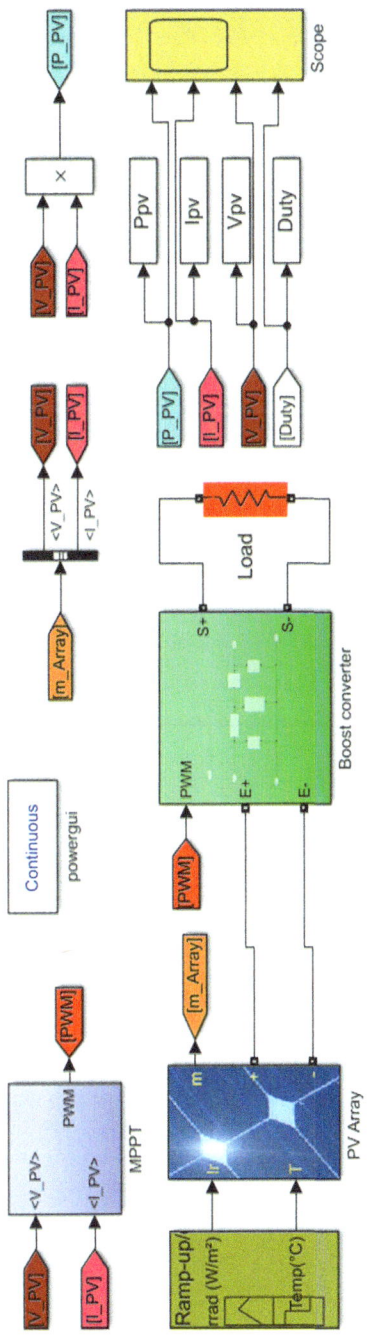

Fig. 4.21 Simulink implementation of the complete standalone PV system

Fig. 4.22 Solar irradiance profiles: **a** sudden insolation changes and **b** sinusoidal insolation changes

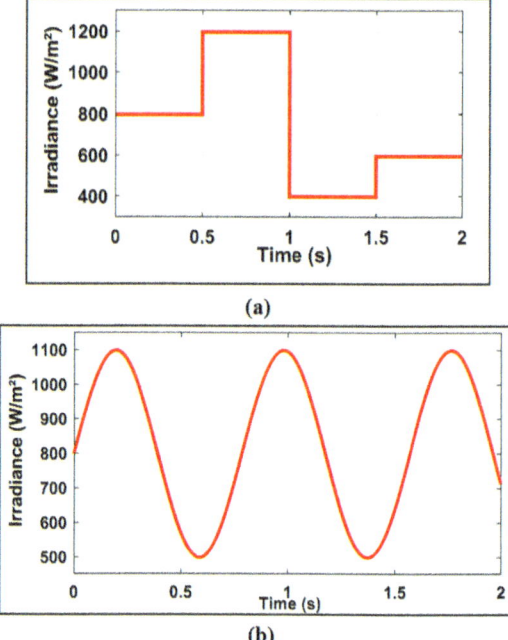

4.3.2.2 Simulation Results in Presence of Load Varying

To further assess the performance of the proposed MPPT approach, a load variation test was conducted. The objective of this test was to evaluate the proposed method's performance in the presence of sudden changes in the load profile. As shown in Fig. 4.31, the load was abruptly increased from 10 to 30 Ω at 0.03 s and then decreased to 20 Ω at 0.06 s, with solar irradiance and temperature kept at STC.

The PV power and duty ratio curves obtained during the load variation test are shown in Figs. 4.32 and 4.33, respectively. The results clearly indicate that the newly proposed MPPT strategy exhibits robust tracking capabilities in response to sudden changes in load conditions. It effectively tracks the expected MPP with higher speed convergence, reduced tracking time (0.007 s), and minimal oscillations, as demonstrated in the zoomed-in view of Fig. 4.32. This superior tracking performance is also evident in the duty cycle curve shown in Fig. 4.33, which adapts accurately with load changes while maintaining optimal photovoltaic power extraction.

Conversely, while the INC and P&O methods were able to follow the MPP for most of the test duration, their performance was negatively affected by load changes, as indicated by the presence of overshoots in the PV power curves and significant oscillations due to unstable duty cycles, as depicted in Fig. 4.33. Furthermore, the FLC method completely lost its ability to track the MPP following the initial load change.

4.3 Second Proposed Algorithm: A Novel MPPT Tactic with Fast ...

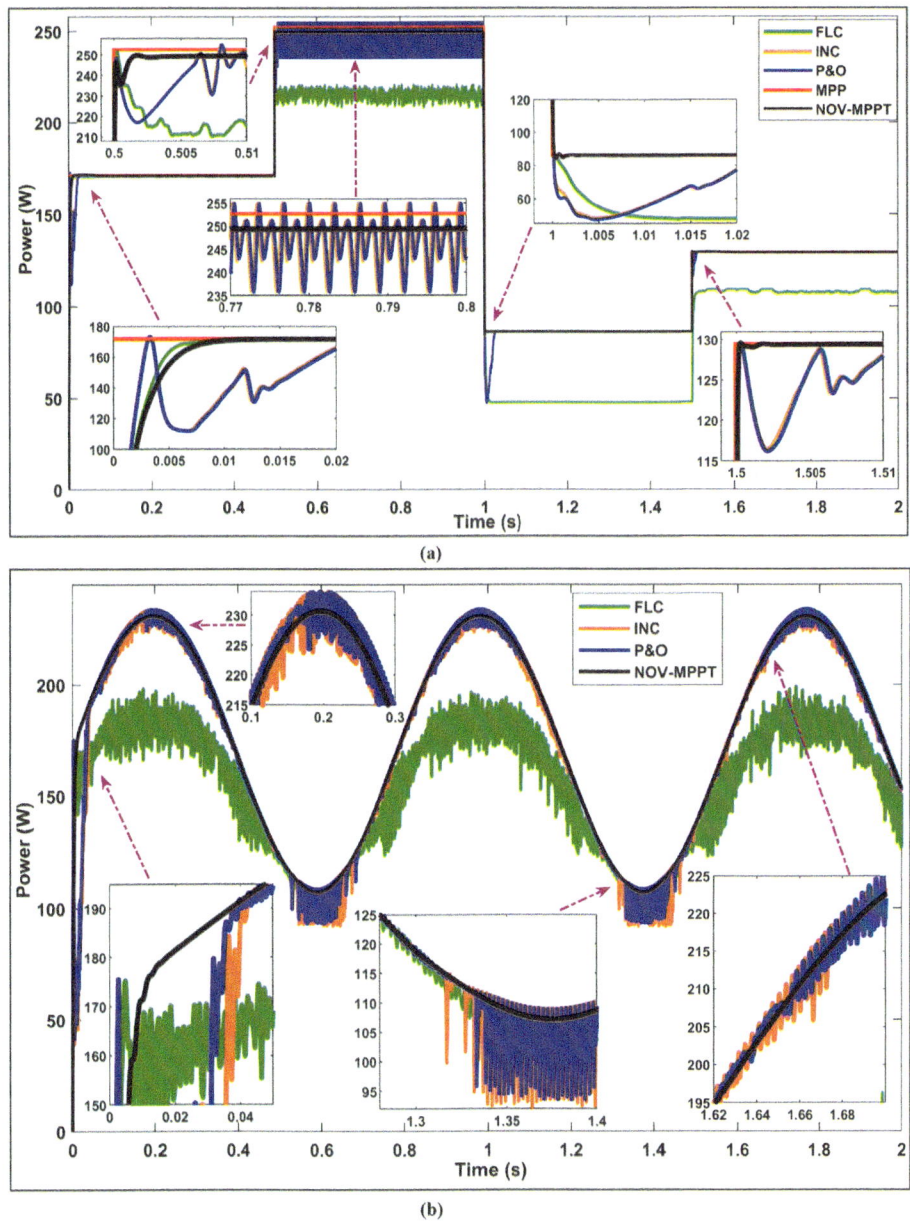

Fig. 4.23 PV power tracked by different MPPT techniques under **a** sudden and **b** sinusoidal test conditions

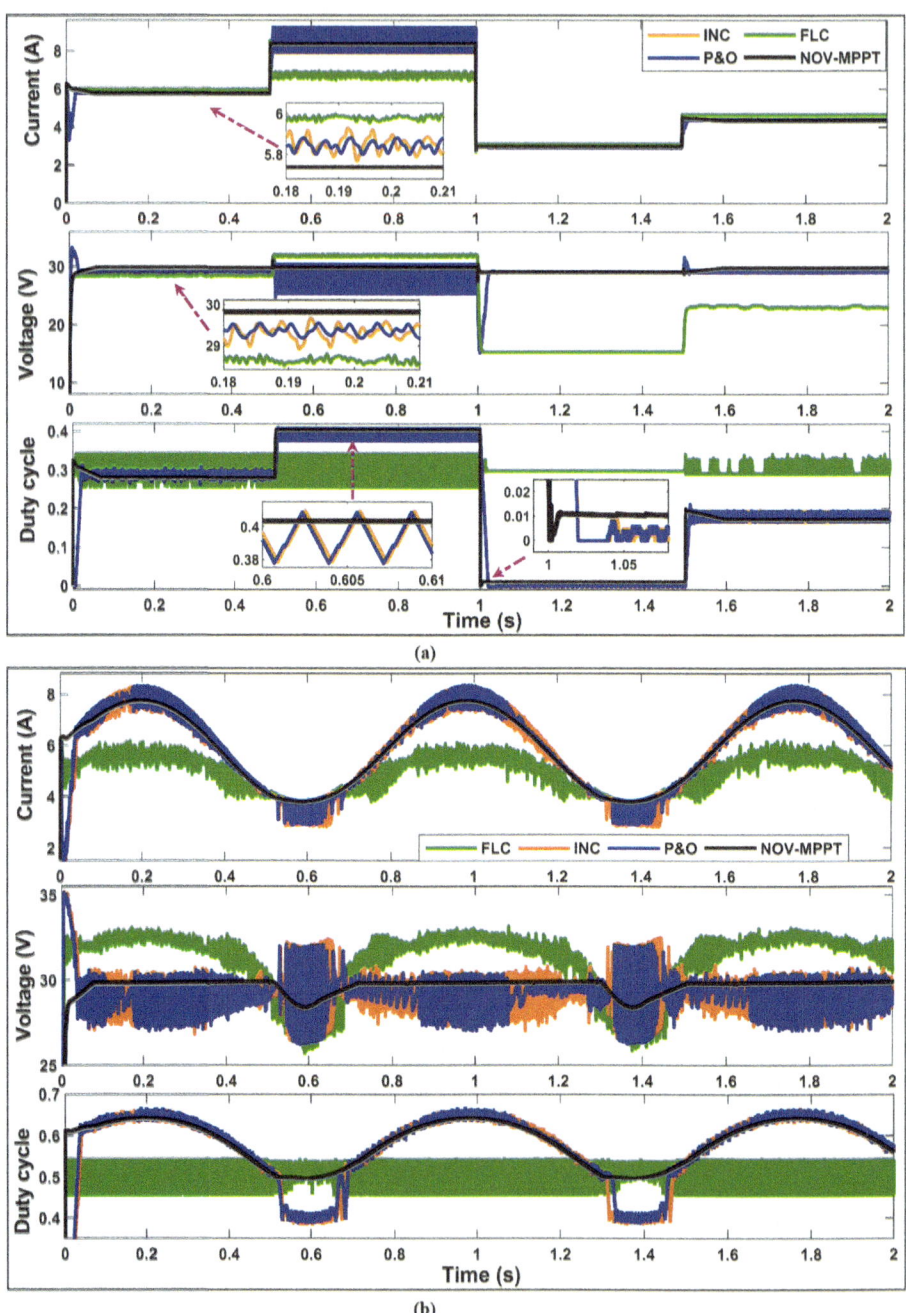

Fig. 4.24 PV current, PV voltage, and duty cycle of different MPPT techniques under **a** sudden and **b** sinusoidal test conditions

4.3 Second Proposed Algorithm: A Novel MPPT Tactic with Fast ...

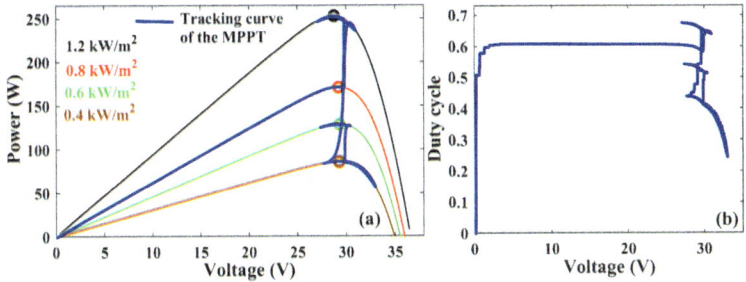

Fig. 4.25 Tracking performance of the proposed approach: **a** P-V curve and **b** D-V curve under sudden insolation changes

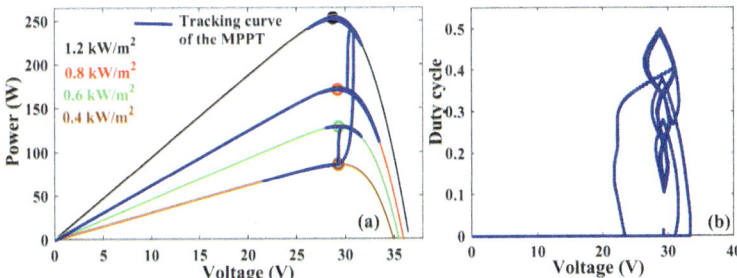

Fig. 4.26 Tracking performance of the P&O method: **a** P-V curve and **b** D-V curve under sudden insolation changes

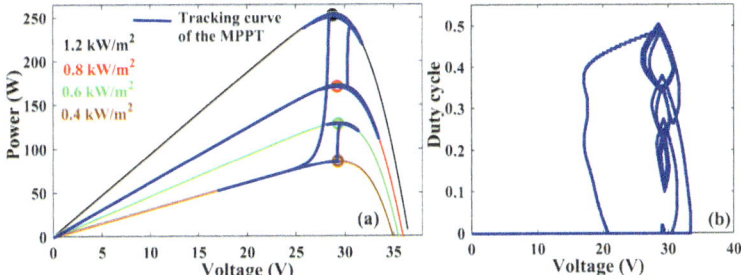

Fig. 4.27 Tracking performance of the INC method: **a** P-V curve and **b** D-V curve under sudden insolation changes

Tables 4.3 and 4.4 present a comparison of the average tracking efficiency and overall performance of the novel MPPT strategy against INC, P&O, and FLC approaches under the MATLAB/Simulink environment. According to these results, the proposed approach

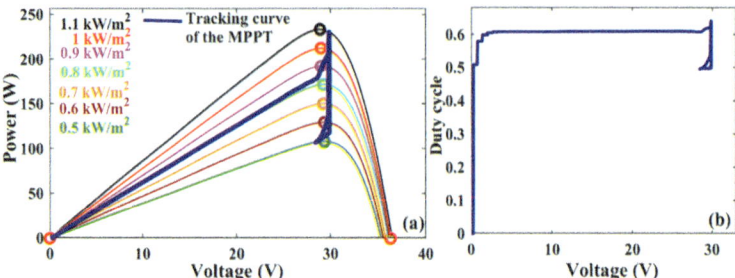

Fig. 4.28 Tracking performance of the proposed approach: **a** P-V curve and **b** D-V curve under sinusoidal insolation changes

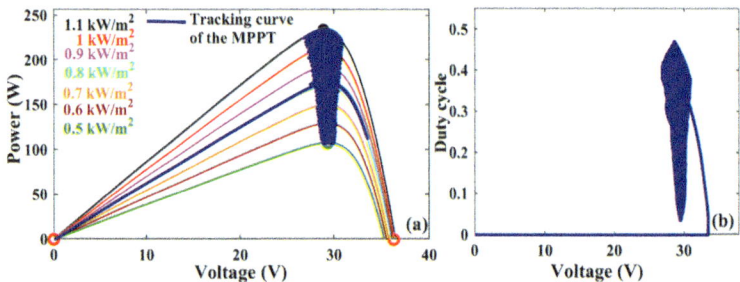

Fig. 4.29 Tracking performance of the P&O method: **a** P-V curve and **b** D-V curve under sinusoidal insolation changes

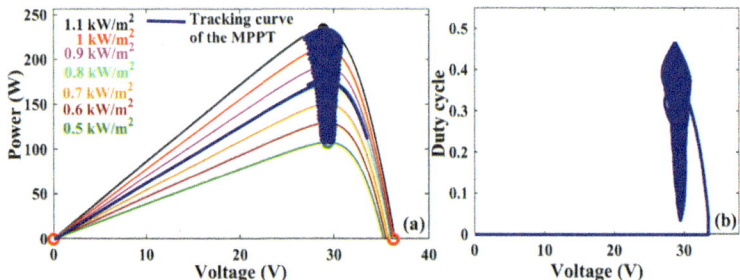

Fig. 4.30 Tracking performance of the INC method: **a** P-V curve and **b** D-V curve under sinusoidal insolation changes

demonstrates significant performance improvements over its contemporaries under all test conditions.

4.3 Second Proposed Algorithm: A Novel MPPT Tactic with Fast ...

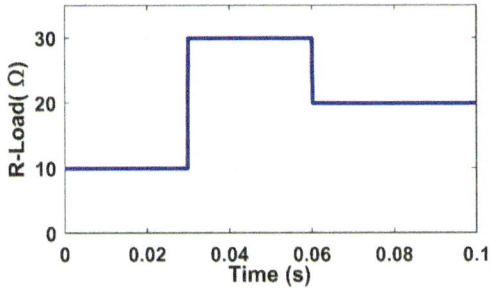

Fig. 4.31 Load profile test under varying load conditions

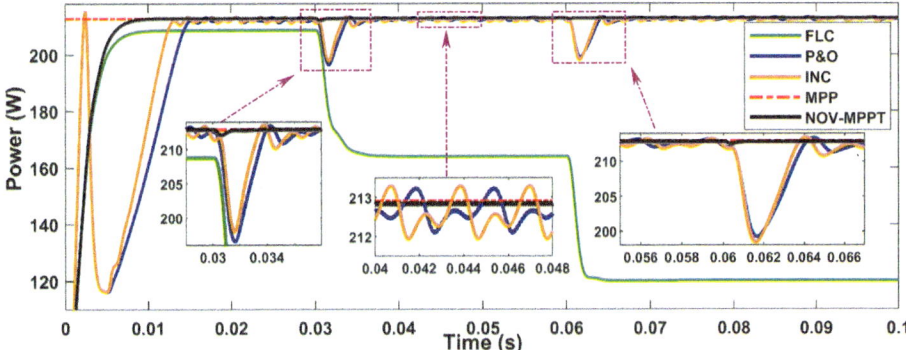

Fig. 4.32 PV power waveforms tracked by MPPT techniques under load variation test conditions

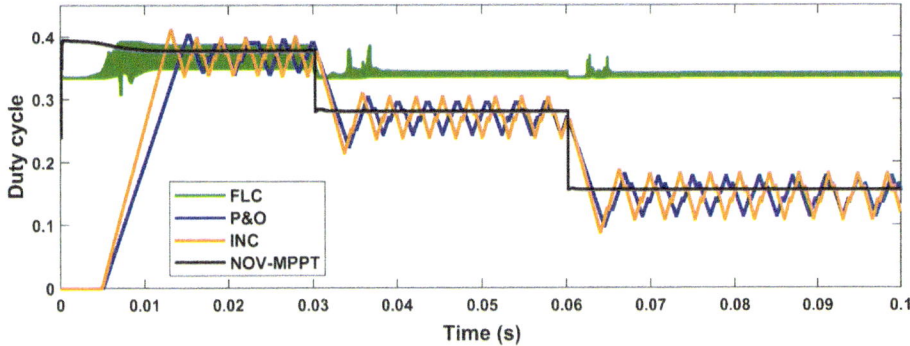

Fig. 4.33 Duty cycle curves of MPPT techniques under load variation test conditions

Table 4.3 Average tracking efficiency comparison of MPPT methods under MATLAB/Simulink environment

	Solar irradiation variation		Load variation
	Sudden test	Sin test	STC conditions
MPPT schemes	Average efficiency (%)		
P&O	99.12	98.52	96.95
INC	99.11	98.32	97.03
FLC	80.98	88.68	77.14
Proposed	99.40	99.57	99.76

Table 4.4 Performance summary of different MPPT strategies based on simulation results

	MPPT schemes			
	P&O	INC	FLC	Proposed
Tracking speed	Slow	Slow	Medium	Faster
Steady-state oscillation	Large	Large	Medium	Zero
Average dynamic efficiency (%)	98.19	98.15	82.26	99.57
Average static efficiency (%)	99.02	99.05	98	99.6
Time response (ms)	14	12	8	7
Power overshoot	High	High	Huge	Insignificant

4.3.3 Implementation and Simulation Results in Proteus Environment

Proteus software is a popular simulation environment for PV systems due to its ease of use and extensive library of electronic components. One of the key components available is the ATMega328 microcontroller, used with the Arduino UNO board. This combination is used to implement an MPPT controller for a standalone PV system, as shown in Fig. 4.34.

Figure 4.34 illustrates the implementation of a standalone PV system using Proteus, which includes the MPPT controller using Arduino. The setup allows easy testing of various MPPT algorithms.

Two cases of insolation variations are used to assess the robustness of the newly proposed MPPT approach compared to traditional INC and P&O methods. These insolation profiles are depicted in Fig. 4.35.

- **Case 1**: Insolation changes in steps between 1.2, 1, 0.6, and 0.4 kW/m^2.
- **Case 2**: Insolation changes suddenly from a high level (1.2 kW/m^2) to a lower level (0.4 kW/m^2).

4.3 Second Proposed Algorithm: A Novel MPPT Tactic with Fast …

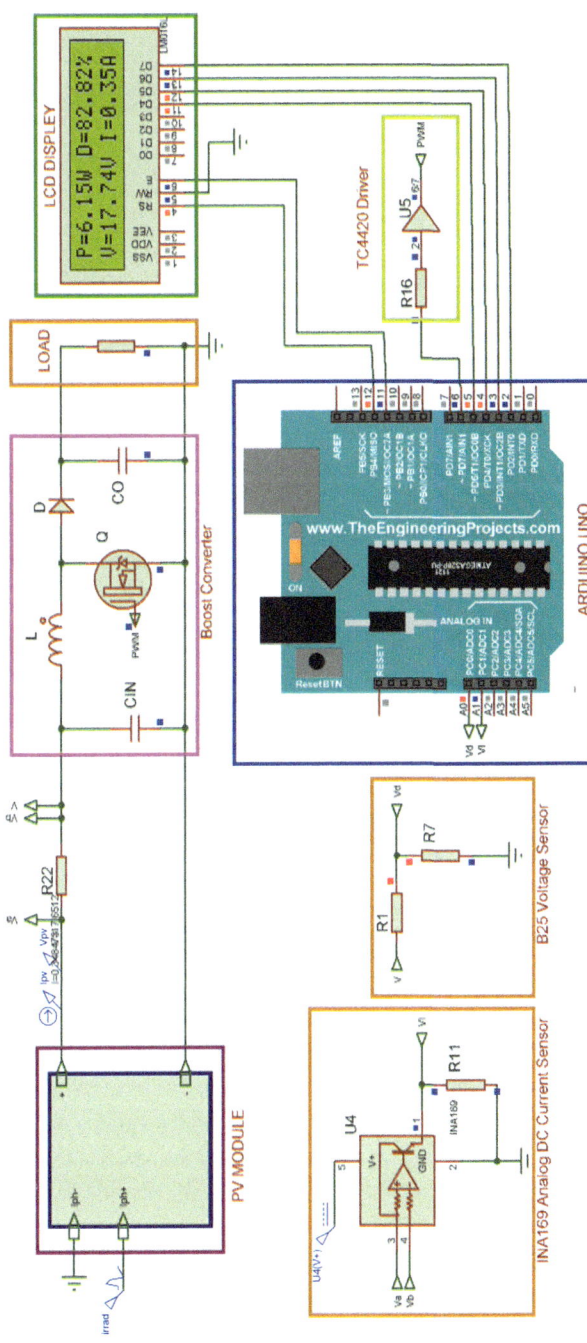

Fig. 4.34 Implementation of the entire standalone PV system using the Proteus environment

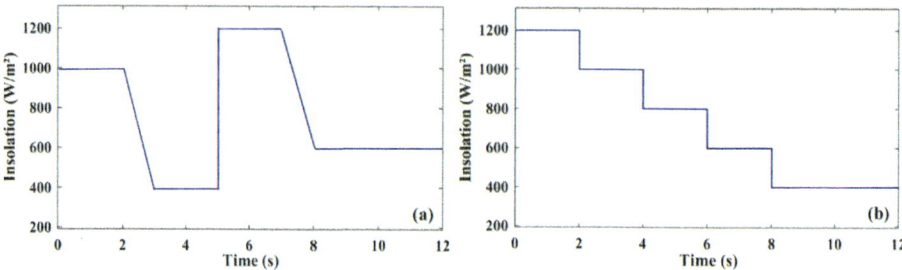

Fig. 4.35 Insolation profile; **a** case 1 and **b** case 2

Figure 4.35a, b illustrate these variations respectively.

The simulation results of photovoltaic power output using different MPPT approaches are shown in Fig. 4.36. The results clearly demonstrate that:

- The newly proposed technique tracks the maximum power point accurately and efficiently in both insolation cases, achieving fast convergence even during sudden changes in insolation. It also minimizes steady-state ripples and avoids sluggish tracking behavior.
- Traditional INC and P&O Techniques suffer from slower response times, noticeable steady-state ripple, and less efficient tracking, especially during sudden insolation variations.

Figure 4.36a shows the power output under **Case 1**, while Fig. 4.36b shows the power output under **Case 2**. The new approach significantly outperforms both traditional methods in terms of tracking accuracy and speed.

4.4 Third Proposed Algorithm: A Novel Adaptable Step Size Theta Approach (ASSTA)

The Adaptable Step Size Theta Approach (ASSTA) is a newly proposed MPPT algorithm specifically designed to address the challenges posed by the non-linearity of PV systems [16]. In PV systems, the MPP is the unique point where the system delivers optimum power, and it is highly sensitive to changing weather conditions, such as variations in temperature and solar irradiance. This behavior is illustrated in the P-V characteristics of the PV array shown in Fig. 4.37:

- Figure 4.37a, b show how the P-V characteristics change under different levels of temperature and solar irradiance.

4.4 Third Proposed Algorithm: A Novel Adaptable Step Size Theta ...

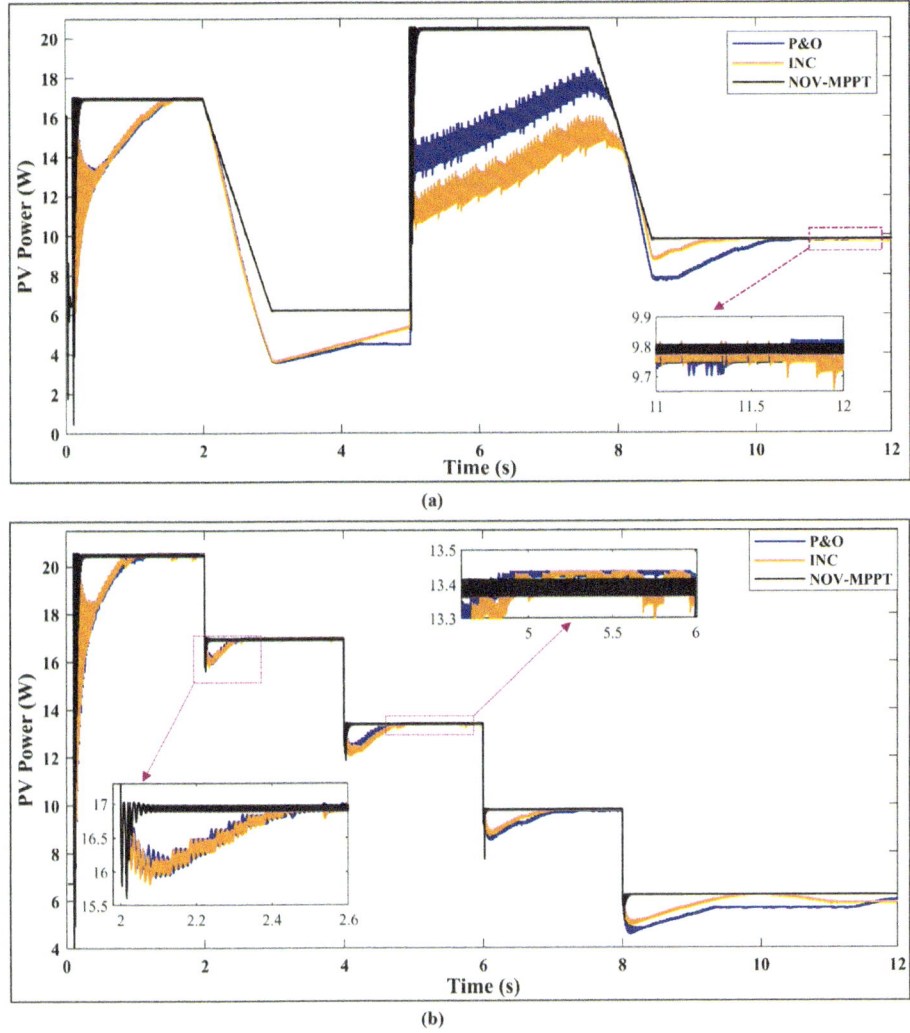

Fig. 4.36 PV power waveforms of MPPT techniques under variation of insolation conditions; **a** case 1 and **b** case 2

- Figure 4.37c illustrates a case where the P-V characteristics present multiple peaks due to non-uniform insolation, often referred to as PSC.

The presence of multiple peaks under PSC makes it challenging for conventional MPPT algorithms to find the correct MPP. Furthermore, rapidly changing atmospheric conditions—such as sudden changes in temperature and solar irradiance—further complicate the task of tracking the available MPP.

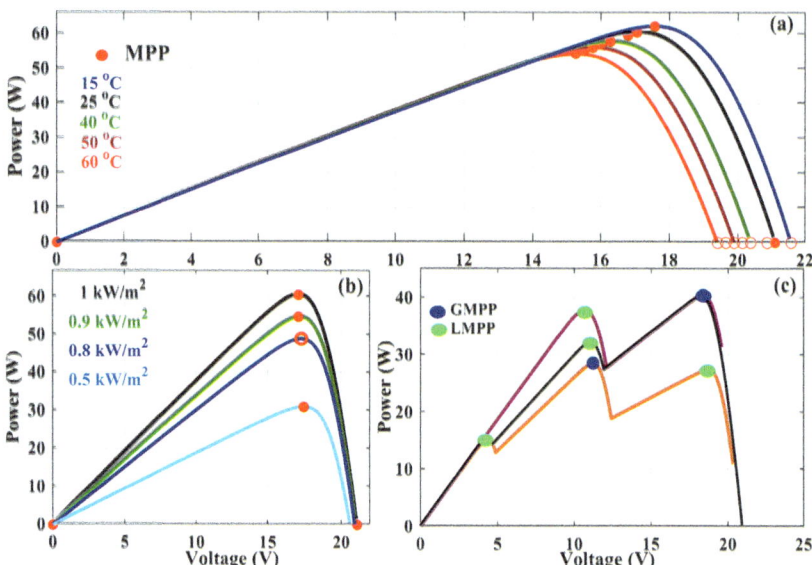

Fig. 4.37 P-V characteristics under change in weather conditions; **a** temperature and **b** solar irradiance and **c** PSC

To address these challenges, the ASSTA is proposed, an adjustable step size-based MPPT strategy designed to:

- Adjust the step size dynamically to efficiently track the MPP, particularly under rapidly changing environmental conditions.
- Solve the inherent problems faced by existing MPPT algorithms, such as sluggish response, high steady-state error, and poor adaptability in variable conditions.
- Improve tracking performance by reliably differentiating between local and global MPPs, ensuring the correct MPP is always tracked.

By employing the adaptable step size, ASSTA aims to provide faster and more accurate tracking of the MPP, resulting in better performance, especially in challenging conditions such as partially shaded situations and rapid atmospheric changes.

4.4.1 Principle

The proposed ASSTA is designed to efficiently track the MPP of a PV system, even under varying environmental conditions. This method relies on the Theta (θ) angle, which is derived from the arctangent of the rate of change of power with respect to voltage, denoted mathematically as arctan (dP_{pv}/dV_{pv}) in Eq. (4.10). The concept of the

4.4 Third Proposed Algorithm: A Novel Adaptable Step Size Theta ...

θ angle and its corresponding details are graphically represented in Fig. 4.38, providing a visual understanding of its role in MPP tracking. In terms of mathematical background, the current and power of an ideal PV cell are represented by Eqs. (4.11) and (4.12), respectively.

$$\theta = \arctan\left(\frac{dP_{pv}}{dV_{pv}}\right) \quad (4.10)$$

$$I_{pv} = I_{ph} - I_0\left[\exp\left(\frac{q \times V_{pv}}{A \times K \times T}\right) - 1\right] \quad (4.11)$$

Similarly, the PV power (Ppv) is the product of voltage and current, expressed as:

$$P_{pv} = V_{pv} \times I_{pv} = V_{pv} \times \left[I_{ph} - I_0\left(\exp\left(\frac{q \times V_{pv}}{A \times K \times T}\right) - 1\right)\right] \quad (4.12)$$

To determine the slope of the power-to-voltage (P-V) curve (shown in Fig. 4.38), the derivative of power with respect to voltage ($dPpv/dVpv$) is derived, as shown in Eq. (4.13). This slope forms the basis for calculating the θ angle. The θ angle itself is defined as the arctangent of this slope, expressed by Eq. (4.14) [16].

$$\frac{dP_{pv}}{dV_{pv}} = I_{pv} - \frac{q \times V_{pv} \times I_0}{A \times K \times T} \exp\left(\frac{q \times V_{pv}}{A \times K \times T}\right) \quad (4.13)$$

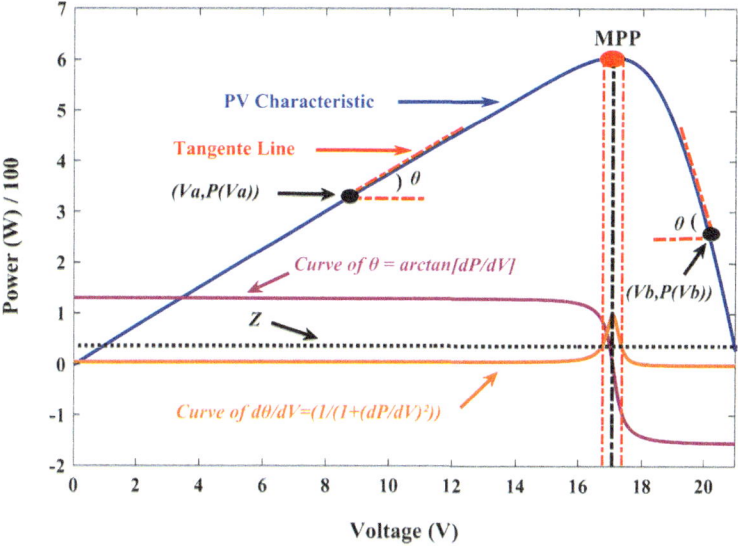

Fig. 4.38 The geometry of the Theta (θ) angle and its derivation $d\theta/dVpv$ diagram in the P-V characteristics under normal climatic conditions

$$\theta = \arctan\left(\frac{dP_{pv}}{dV_{pv}}\right) = \arctan\left[I_{pv} - \frac{q \times V_{pv} \times I_0}{A \times K \times T} \exp\left(\frac{q \times V_{pv}}{A \times K \times T}\right)\right] \quad (4.14)$$

Additionally, the derivative of the Theta angle with respect to voltage ($d\theta/dV_{pv}$) is obtained as shown in Eq. (4.15) [16]:

$$\frac{d\theta}{dV_{pv}} = \frac{d}{dV_{pv}}\left[\arctan\left(\frac{dP_{pv}}{dV_{pv}}\right)\right] = \frac{1}{1 + \left\{\frac{dP_{pv}}{dV_{pv}}\right\}^2} = \frac{1}{1 + \left\{I_{pv} - \frac{q \times V_{pv} \times I_0}{A \times K \times T} \exp\left(\frac{q \times V_{pv}}{A \times K \times T}\right)\right\}^2} \quad (4.15)$$

The θ angle has three possible values based on the position of the OP in relation to the MPP: $\theta = 0$ when the OP is at the MPP, $\theta > 0$ when the OP is on the left side of the MPP, and $\theta < 0$ when the OP is on the right side of the MPP, as indicated by Eq. (4.16). The ASSTA algorithm incorporates an adjustable step size (M) (as shown in 4.39) to enhance the tracking performance of the MPP, with M being calculated, as given in Eq. (4.17). This adjustable step size is then used to dynamically adjust the duty cycle ($d(k)$), where SF is a scaling factor between 0 and 1 that helps strike a balance between dynamic response and ripple reduction, as shown in Eq. (4.18). The duty cycle value varies depending on the OP's distance from the MPP: it is high when the OP is far from the MPP (either left or right), small when close to the MPP, and null when precisely at the MPP (Fig. 4.40).

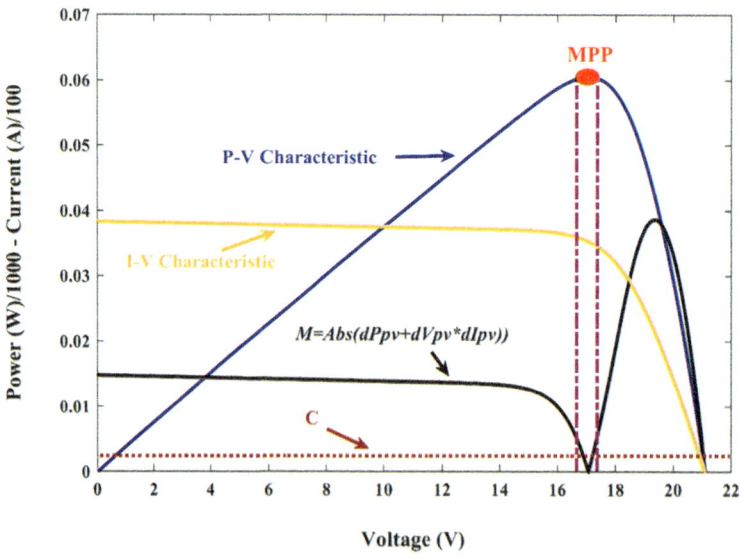

Fig. 4.39 Characteristics of P-V, I-V and $(dPpv + dVpv * dIpv)$-V under uniform climatic conditions

4.4 Third Proposed Algorithm: A Novel Adaptable Step Size Theta ...

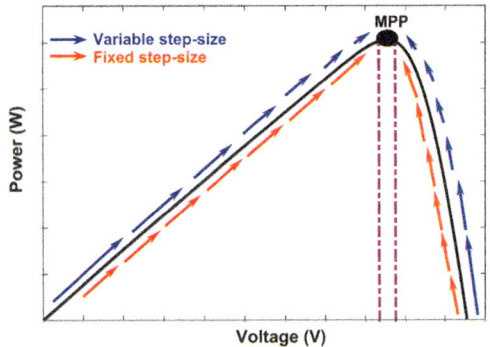

Fig. 4.40 Graphical illustration of MPPT operation in case of using a fixed and adjustable perturbation step size

The flowchart of the ASSTA algorithm, depicted in Fig. 4.41, outlines the sequence of operations. It begins by measuring the voltage and current of the PV array. Next, M is calculated. If M is less than a predefined constant C ($M < C = 0.003$), the duty cycle remains unchanged. The θ angle and its derivative ($d\theta/dVpv$) are then checked: if $\theta = 0$ and $d\theta/dVpv > Z$ (where $Z = 0.02$), the duty cycle remains constant, as this indicates the system has reached the MPP zone. If these conditions are not satisfied, the algorithm determines whether the OP is to the left or right of the MPP and adjusts the duty cycle accordingly, using Eq. (4.18). This strategic approach significantly enhances tracking speed, improves dynamic performance, and reduces oscillations around the MPP, ensuring efficient tracking even under rapidly changing environmental conditions.

$$\begin{cases} \theta = 0 & \text{At MPP} \\ \theta > 0 & \text{At the left side of MPP} \\ \theta < 0 & \text{At the right side of MPP} \end{cases} \quad (4.16)$$

$$M = abs(dP_{pv} + dV_{pv} \times dI_{pv}) \quad (4.17)$$

$$d(k) = d(k-1) \pm SF \times M \quad (4.18)$$

4.4.2 MATLAB/Simulink Implementation and Simulation Results

The implementation and simulation of the entire standalone PV system were conducted using the Simulink environment in MATLAB, as depicted in Fig. 4.42. The parameters of the PV module and the DC-DC boost converter used in the simulation are detailed in Table 4.5. To evaluate the accuracy, performance, and robustness of the newly developed ASSTA, four distinct simulation scenarios were designed to simulate varying weather conditions. In Scenario 1, STC were applied with an insolation of 1 kW/m² and a temperature of 25 °C, referred to as Case 1. Scenario 2 involved an abrupt change in insolation,

Fig. 4.41 Flow chart of the novel ASSTA MPPT approach

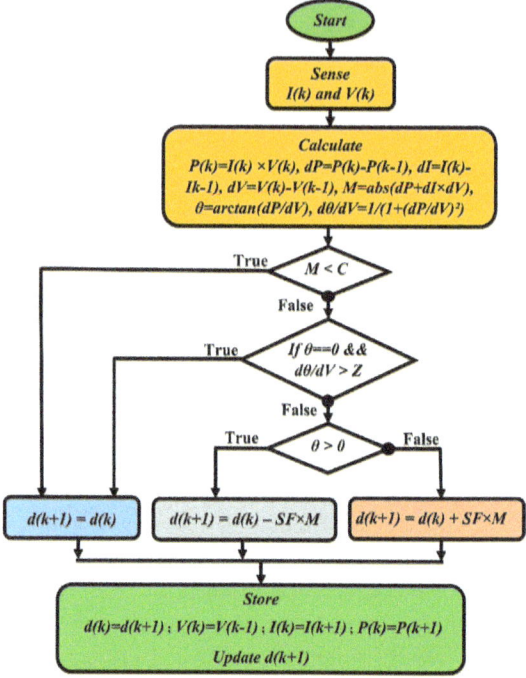

with sudden variations in solar irradiance, while the temperature was kept fixed at 25 °C. Scenario 3 introduced an abrupt change in temperature, where the temperature was varied while the insolation was kept constant at 1 kW/m². Scenario 4 simulated PSC, performed under two different shading scenarios, referred to as Case 2 and Case 3, as depicted in Fig. 4.43. To comprehensively evaluate the effectiveness of the ASSTA approach, it was compared against several well-known MPPT techniques commonly used in PV systems. These included traditional methods such as INC and P&O, an improved approach like the Variable Step Size Incremental Conductance (VSSINC), and a metaheuristic algorithm, the PSO MPPT method. This comparison was conducted to benchmark ASSTA's performance under different conditions, including abrupt changes in insolation and temperature as well as partial shading scenarios. The results provide valuable insights into how ASSTA performs in real-world, dynamic environments compared to these established MPPT techniques.

4.4.2.1 Simulation Results According to STC and Sudden Change in Solar Irradiation Conditions

The simulation results for the PV power and duty cycle waveforms using the novel Adaptable Step ASSTA were thoroughly compared with those obtained using traditional and

4.4 Third Proposed Algorithm: A Novel Adaptable Step Size Theta ...

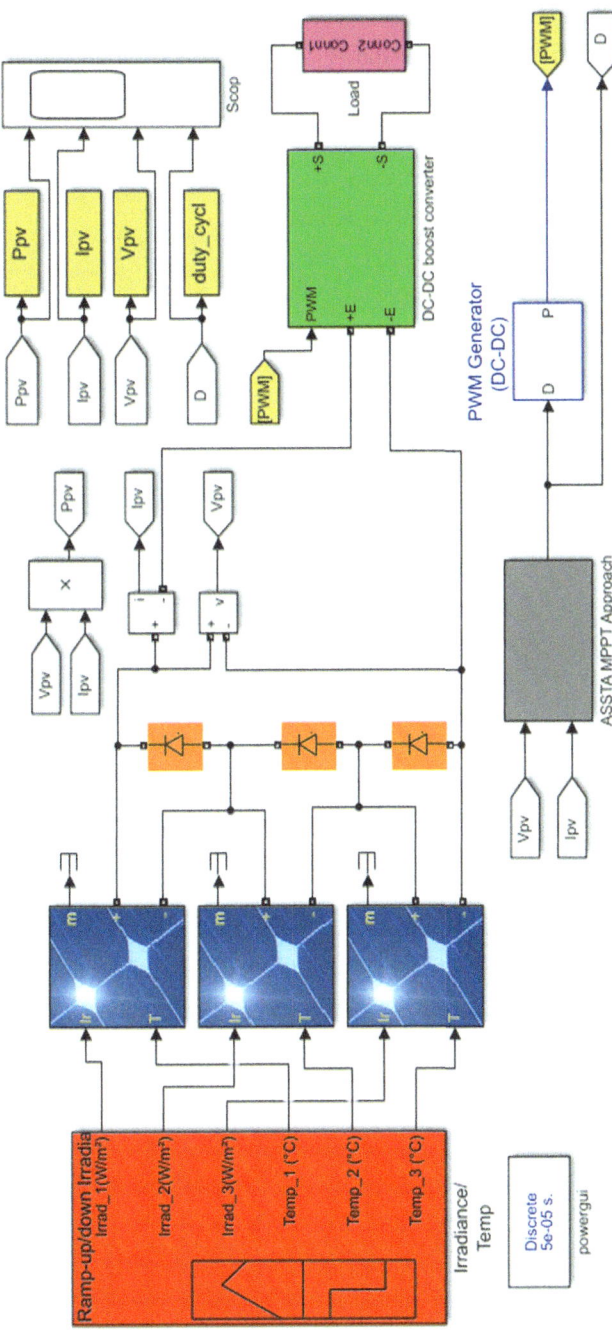

Fig. 4.42 The entire PV system implementation using the Simulink environment of MATLAB software

Table 4.5 Parameters of the complete photovoltaic system (MSX-60 PV module) in the STC

Parameters	Variable	Value
PV module		
Power at MPP	P_{MPP}	60.50 W
MPP voltage	V_{MPP}	17.04 V
MPP current	I_{MPP}	3.55 A
Voltage of open circuit	V_{oc}	21.09 V
Current of short circuit	I_{sc}	3.8 A
DC-DC boost converter		
Input capacitor	C_{in}	100 µF
Output capacitor	C_{out}	100 µF
Inductor	L	3 mH
Switch Frequency	F	10 kHz
Resistive Load	R_{load}	30 Ω

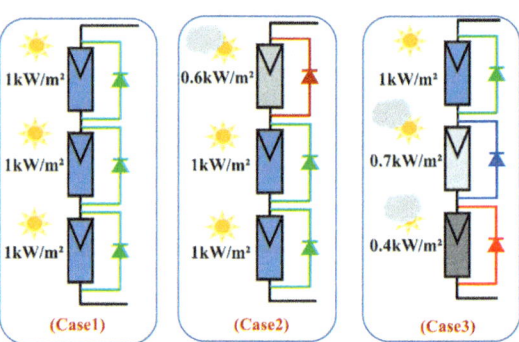

Fig. 4.43 PV array configurations

metaheuristic MPPT techniques, including INC, P&O, VSSINC, and PSO. These comparisons were made under both STC and dynamic scenarios involving sudden insolation changes from 0.5 to 1 kW/m² and then down to 0.7 kW/m², as depicted in Fig. 4.44a, b.

In Fig. 4.44a, which illustrates the PV power response over time, ASSTA (blue line) shows a stable, consistent approach to the MPP with minimal oscillations, even under dynamic insolation changes. The zoomed-in views reveal ASSTA's precise power tracking with almost no fluctuation around the MPP, indicating high efficiency and stability.

On the other hand, INC (black) and P&O (red) exhibit substantial oscillations around the MPP, particularly evident in the close-up sections, where they fail to converge smoothly. This oscillatory behavior around the MPP in INC and P&O methods results in inefficient power tracking, especially noticeable during sudden changes in insolation, as shown in the figure's zoomed-in insets.

4.4 Third Proposed Algorithm: A Novel Adaptable Step Size Theta ...

While the PSO technique (green) performs adequately under stable conditions, it struggles to adapt quickly to the abrupt changes in insolation. This limitation results in delayed responses, reducing overall tracking efficiency in dynamic environments. VSS-INC, while showing faster convergence than INC and P&O, still demonstrates slight inconsistencies around the MPP, though with a more stable output than PSO.

The duty cycle curves in Fig. 4.44b highlight ASSTA's responsiveness and adaptability to environmental changes. ASSTA's duty cycle stabilizes quickly with minimal oscillations, allowing the PV system to maintain optimal power levels without excessive adjustments. The figure's zoomed-in sections illustrate ASSTA's smooth, controlled transitions, reinforcing its tracking precision even during abrupt shifts in insolation.

In contrast, the INC and P&O methods exhibit significant fluctuations in their duty cycles, which results in erratic power outputs, as previously noted. The VSSINC technique, while faster in its duty cycle response than INC and P&O, still displays occasional instability, impacting its tracking consistency. PSO's duty cycle exhibits delayed adaptation, especially noticeable in dynamic conditions, leading to inefficient power extraction.

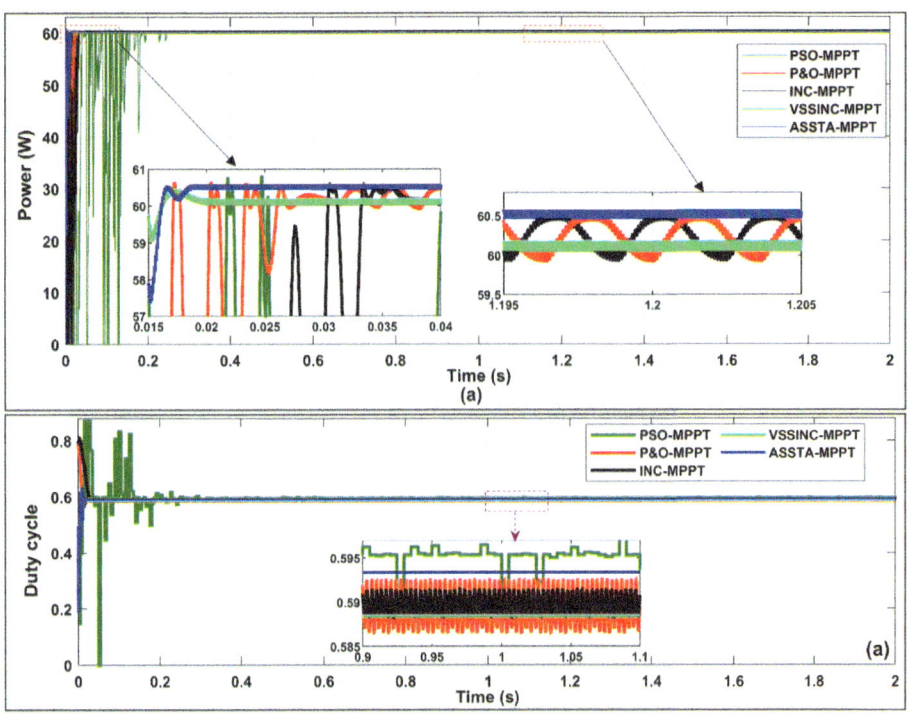

Fig. 4.44 Comparison of the PV power and duty cycle waveforms of the novel ASSTA with that of the INC, P&O, VSSINC and PSO MPPT strategies under; **a** STC and **b** sudden insolation variation conditions

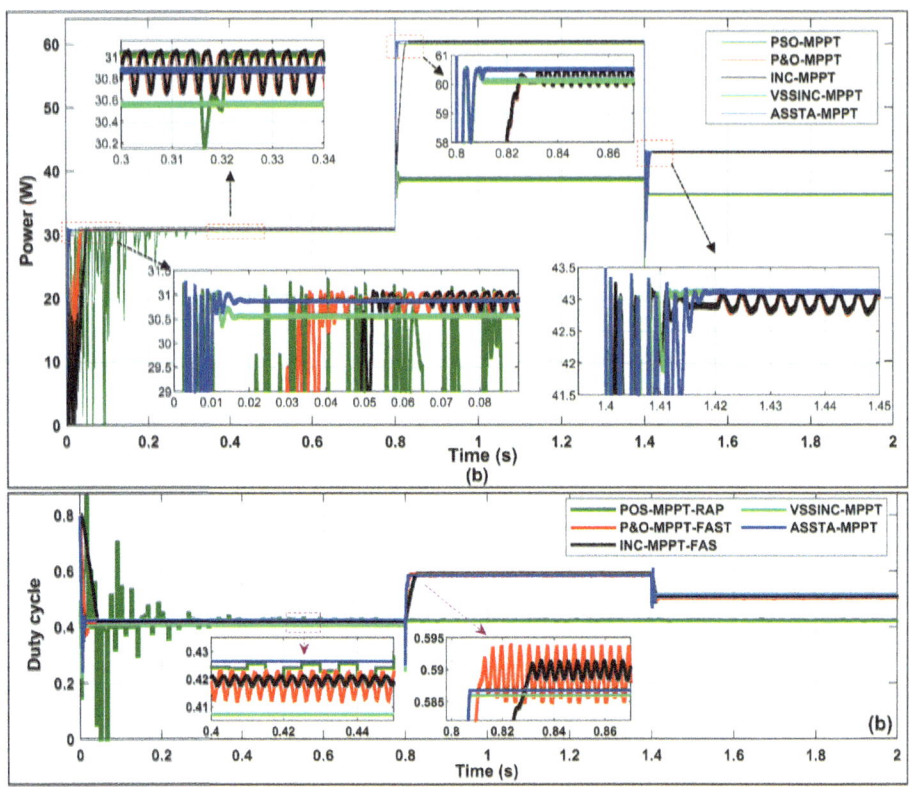

Fig. 4.44 (continued)

The results indicate that the ASSTA approach consistently outperforms other MPPT techniques across both STC and dynamic insolation conditions. ASSTA's duty cycle and PV power curves show minimal oscillations and rapid convergence, achieving tracking efficiencies of 99.84% under STC and 99.55% under fluctuating insolation. This high level of robustness enables ASSTA to quickly adapt to environmental variations, making it a reliable choice for efficient MPP tracking in photovoltaic systems.

In conclusion, the novel ASSTA approach surpasses traditional and metaheuristic MPPT methods, especially under rapidly changing conditions. Its superior tracking accuracy, fast response, and stability ensure maximized power output from PV systems, making ASSTA a highly effective solution for optimizing energy production in dynamic environments.

4.4.2.2 Simulation Results According to Sudden Change in Temperature Conditions

Figure 4.45 presents the simulation results comparing the PV power and duty cycle curves for the novel ASSTA approach against INC, P&O, VSSINC, and PSO methods under conditions of sudden temperature changes.

In this scenario, the temperature increased abruptly from 25 to 40 °C at 0.4 s, then to 50 °C at 0.8 s, before dropping to 35 °C at 1.2 s, and finally settling at 40 °C at 1.6 s. The results demonstrate that the novel ASSTA method closely tracks the MPP throughout the temperature variations, exhibiting superior tracking performance with minimal power loss and a rapid response time.

In Fig. 4.45a, the PV power waveform clearly illustrates ASSTA's smooth and efficient behavior, effectively minimizing oscillations during rapid temperature shifts. ASSTA's

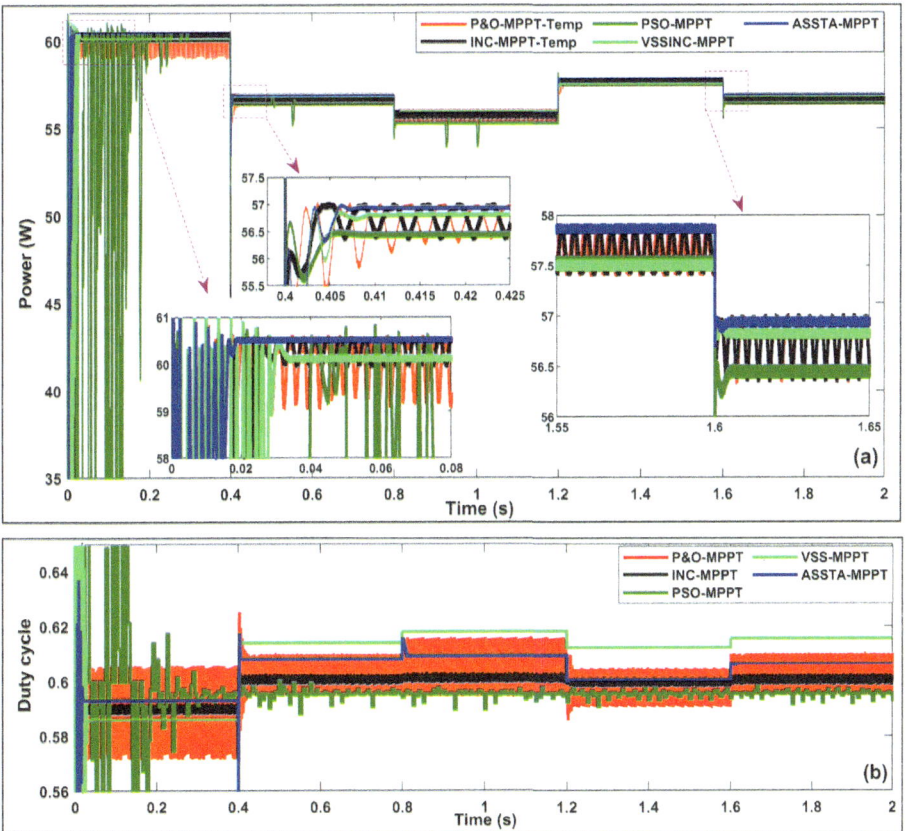

Fig. 4.45 Comparison of the novel ASSTA with INC, P&O, VSSINC, and PSO strategies under fast temperature change scenario; **a** PV power, and **b** duty cycle waveforms

fast response is attributed to its dynamically adaptive duty ratio, which handles sudden changes in temperature, as shown in Fig. 4.45b.

In contrast, the PSO method experienced pronounced oscillations at the initial temperature transition and failed to converge properly to the MPP beyond 0.3 s, resulting in poor tracking. Both INC and P&O approaches also exhibited significant oscillations, particularly in P&O, which hampered their ability to track the MPP effectively. Although the VSSINC method successfully reduced oscillations compared to P&O and INC, it still struggled to achieve precise MPP tracking, as evidenced in the zoomed-in sections of Fig. 4.45a.

The average tracking efficiency of ASSTA reached 99.81%, significantly outperforming the other techniques. Specifically, PSO achieved 96.73%, while VSSINC, INC, and P&O recorded efficiencies of 99.33, 99.23, and 99.18%, respectively. These results highlight ASSTA's superior capability to maintain high tracking efficiency while minimizing oscillations, even in scenarios involving rapid and considerable temperature changes.

4.4.2.3 Simulation Results According to PSC

Figure 4.46 illustrates the comparative performance of the PV power output and duty ratio waveforms for the proposed ASSTA, along with INC, P&O, VSSINC, and PSO MPPT strategies under PSC for Case 2 and Case 3.

The results show that the INC, P&O, and VSSINC methods struggle to track the GMPP and instead settle at LMPP, indicating limited ability to navigate the multi-peak characteristic of the power-voltage (P-V) curve under non-uniform irradiance. On the other hand, the PSO algorithm shows better capability in locating the GMPP but suffers from high oscillations and extended convergence times, which detract from its efficiency and stability.

In contrast, the proposed ASSTA approach effectively tracks the GMPP with rapid convergence and minimal fluctuations, demonstrating a marked improvement over the other techniques. ASSTA's adaptive mechanism allows it to efficiently overcome the challenges presented by partial shading, ensuring the PV system consistently operates at the true optimal point to maximize energy yield.

A detailed numerical comparison in Table 4.6 highlights ASSTA's superior performance across various metrics, such as response time, convergence speed, and both static and dynamic tracking efficiency. Under STC, ASSTA achieved a response time of 0.018 s, which is significantly faster than its counterparts. Furthermore, ASSTA exhibited a tracking efficiency of 99.55% under dynamic conditions, surpassing PSO, VSSINC, INC, and P&O, which suffered from slower responses and lower efficiencies.

ASSTA also achieved high tracking efficiency during partial shading scenarios, recording 99.52% in Case 2 and 99.10% in Case 3. In comparison, VSSINC achieved

4.4 Third Proposed Algorithm: A Novel Adaptable Step Size Theta ...

93.33% in Case 2 and 94.85% in Case 3, while P&O and INC struggled to effectively track the global peak, especially under challenging shading patterns.

In terms of mitigating oscillations, ASSTA demonstrated significant robustness, nearly eliminating fluctuations compared to the other methods. INC and P&O exhibited high oscillation levels, and even PSO showed moderate instability. ASSTA's near-complete suppression of these oscillations contributes to reducing energy losses and maximizing PV system output.

In conclusion, the ASSTA approach demonstrates substantial superiority over conventional and metaheuristic MPPT techniques, particularly under PSC or other challenging environmental dynamics. Its fast convergence, precise tracking, high efficiency, and stability make ASSTA a valuable advancement for enhancing the performance of photovoltaic systems. The performance metrics in Table 4.6 clearly support ASSTA's advantages over other approaches.

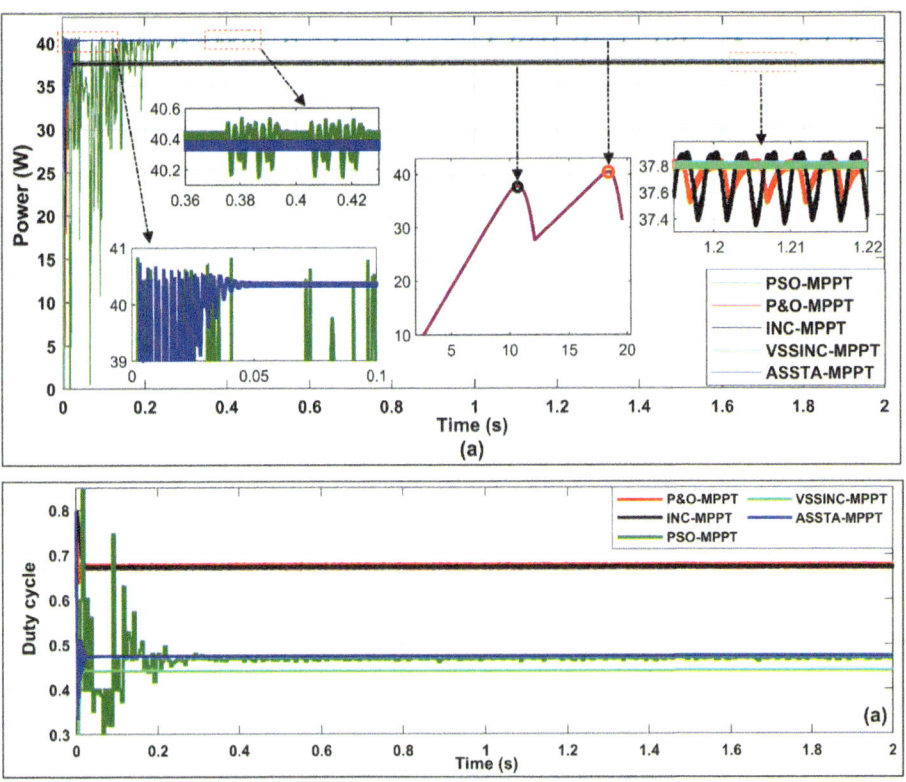

Fig. 4.46 Comparison of PV power and duty ratio waveforms of ASSTA with INC, P&O, VSSINC, and PSO under PSC; **a** Case 2, **b** Case 3

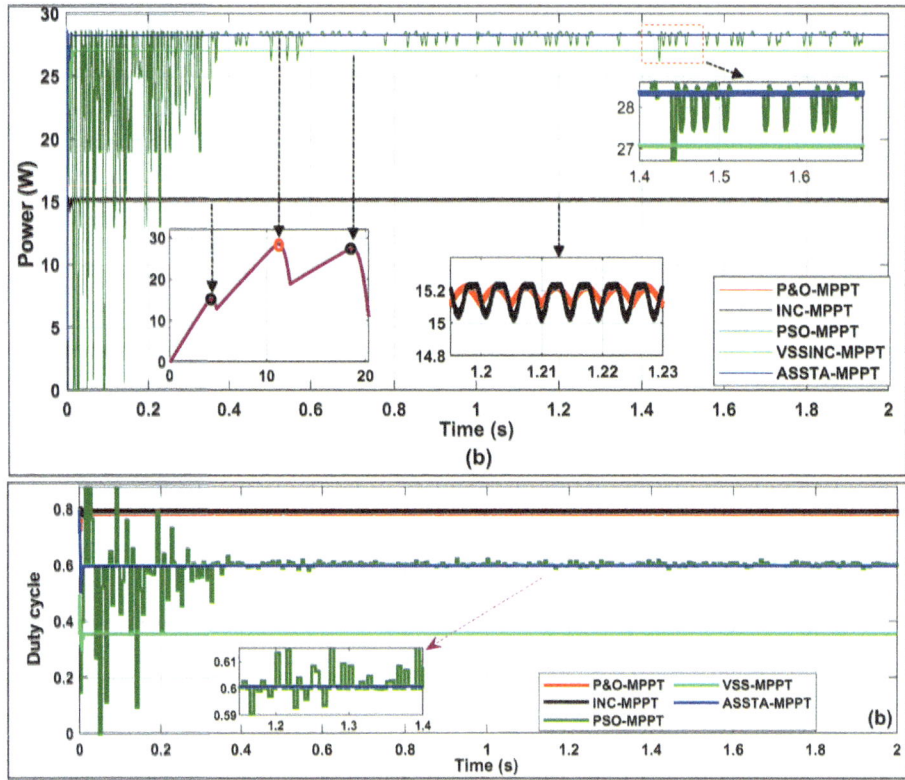

Fig. 4.46 (continued)

4.5 Fourth Proposed Algorithm: An Innovative MPPT Approach for Temperature Varying with Zero Fluctuation and Fast-Converging Speed

4.5.1 Principle

The innovative MPPT strategy proposed here is designed based on observed patterns in the current-power (I-P) and current-voltage (I-V) characteristics of PV modules at different temperature levels. As illustrated in Fig. 4.47, the MPP current of the PV module, under significant temperature variations, remains within a limited range referred to as the MPP Current Zone (CZ). Within this zone, the IMPP fluctuates between two threshold values: the minimum ($I_{MPP_{min}}$) and maximum ($I_{MPP_{max}}$) MPP currents [17].

The flowchart in Fig. 4.48 outlines the operation of the proposed MPPT strategy. The process begins by measuring the PV module's current and voltage. The algorithm then

4.5 Fourth Proposed Algorithm: An Innovative MPPT Approach ...

Table 4.6 Summary of the performance comparison of the ASSTA, P&O, INC, VSSINC, and PSO MPPT techniques under the STC, Case 2, and Case 3 of PSC scenarios

	MPPT approaches				
	P&O	INC	VSSINC	PSO	ASSTA
Response time (s) under STC	0.027	0.033	0.022	0.2	0.018
Speed converging	Medium	Medium	High	Low	Very high
Static efficiency of tracking (%) (STC)	99.07	98.76	99.25	97.46	99.84
Dynamic efficiency of tracking (%)	98.42	97.85	99	82.46	99.55
GMPP tracking	No	No	No	Yes	Yes
Tracking efficiency (%) under Case 2 and Case 3 of PSC	93.08 and 53.2	90.37 and 53.13	93.33 and 94.85	98.16 and 95.92	99.52 and 99.10
Oscillations problem	High	High	Negligible	Moderate	Negligible

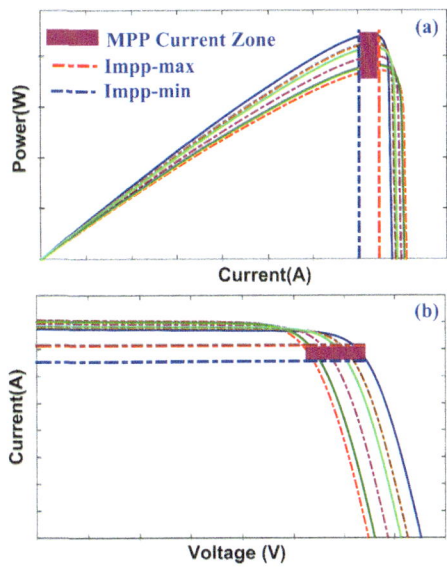

Fig. 4.47 Illustration of the MPP current zone (CZ) in the I-P and I-V characteristics of the PV module

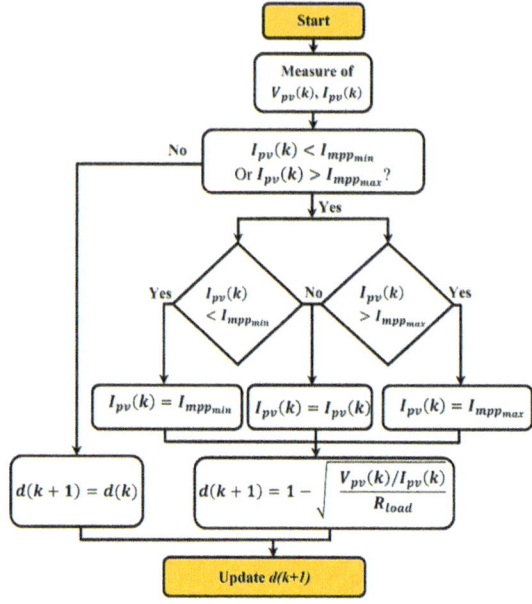

Fig. 4.48 Flow chart of the innovative MPPT scheme

assesses whether the detected current, denoted as $I_{pv}(k)$, falls within the predefined MPP Current Zone. Specifically:

- If the actual current, $I_{pv}(k)$ is less than $I_{MPP_{min}}$, it is assigned the value of $I_{MPP_{min}}$.
- If $I_{pv}(k)$ exceeds $I_{MPP_{max}}$, it is assigned the value of $I_{MPP_{max}}$.

Using these bounds ensures the current remains within the MPP Current Zone, facilitating accurate MPPT performance despite temperature variations.

Following this assessment, the algorithm computes a new duty cycle, $d(k+1)$, based on Eq. (4.8). However, if $I_{pv}(k)$ does not meet the initial condition (i.e., does not fall within the MPP CZ), the new duty cycle, $d(k+1)$, is set equal to the previous duty cycle, $d(k)$.

This method delivers fast tracking time and high convergence speed, achieving reliable MPP tracking under fluctuating temperature conditions without introducing fluctuations in the duty cycle. Consequently, the proposed strategy ensures stable, accurate, and rapid MPPT performance [17].

4.5.2 MATLAB/Simulink Implementation and Simulation Results

To validate the novel MPPT strategy, a series of simulations was conducted under varying temperature and insolation conditions using MATLAB/Simulink. The performance of this

4.5 Fourth Proposed Algorithm: An Innovative MPPT Approach ...

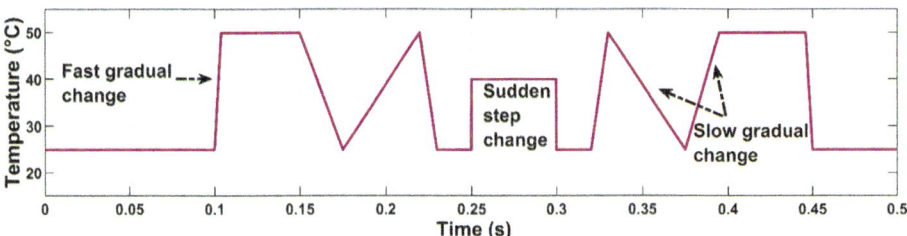

Fig. 4.49 The first scenario of temperature varying

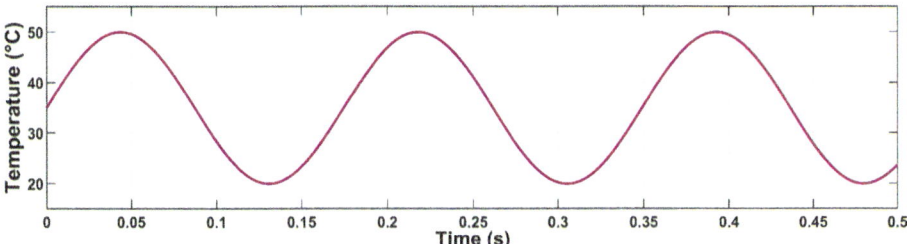

Fig. 4.50 The second scenario of temperature varying

new MPPT method was benchmarked against the INC and P&O MPPT strategies. Two different temperature variation scenarios were tested, as illustrated in Figs. 4.49 and 4.50:

- The first scenario involves gradual temperature changes.
- The second scenario, in a sinusoidal form, represents slow temperature fluctuations.

Each scenario was simulated under two solar irradiance levels: 1 and 0.2 kW/m².

4.5.2.1 Simulation Results According to the First Temperature Condition Scenario Under High and Low Insolation Level

Figures 4.51 and 4.53 present the simulation results for the output power, voltage, current, and duty cycle ratio of the PV system, comparing the proposed MPPT method with the traditional INC and P&O methods. Meanwhile, Figs. 4.52 and 4.54 illustrate the tracking efficiency of the proposed MPPT scheme against the INC and P&O methods under the first temperature scenario, which includes both high and low irradiance conditions.

The results indicate that the proposed MPPT approach demonstrates superior tracking performance. As shown in Figs. 4.51 and 4.53, the proposed method achieves the MPP with high convergence speed, minimal response time, and no oscillations, regardless of temperature fluctuations or varying levels of irradiance. Conversely, the performance of

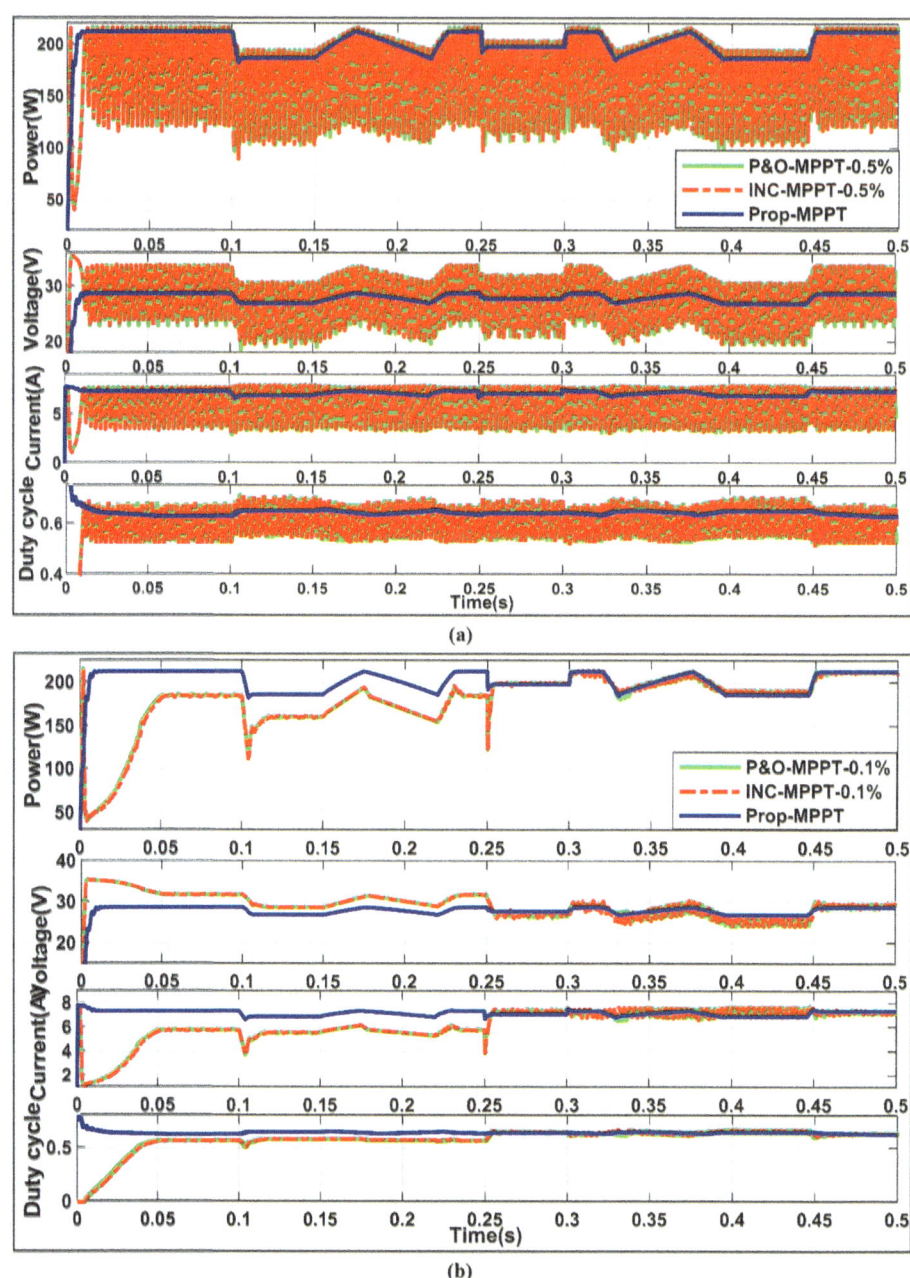

Fig. 4.51 Simulation results comparing the proposed MPPT with INC and P&O under high irradiance (1 kW/m^2) for perturbation step sizes of 0.5 and 0.1%

4.5 Fourth Proposed Algorithm: An Innovative MPPT Approach ...

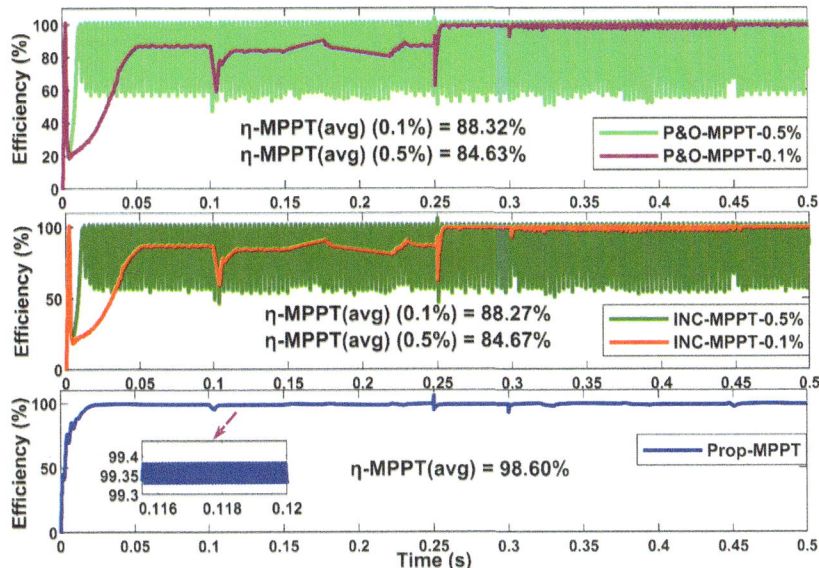

Fig. 4.52 Tracking efficiency of the MPPT methods under high irradiance and the first temperature scenario

the INC and P&O methods is notably affected by temperature changes, leading to significant oscillations around the MPP, especially when a 0.5% perturbation step size is applied. These oscillations in INC and P&O methods, visible in Figs. 4.51a and 4.53a result in increased power loss under high irradiance conditions.

When the perturbation step size is reduced to 0.1%, the INC and P&O methods show further limitations, taking approximately 2.25 s to stabilize at the MPP and experiencing tracking difficulties with each temperature shift. This behavior is particularly evident in Figs. 4.51b and 4.53b, where these methods struggle to maintain MPP under both high and low irradiance.

In terms of tracking efficiency, as illustrated in Figs. 4.52 and 4.54 under the first temperature scenario (with irradiance levels of 1 and 0.2 kW/m^2), the proposed MPPT method exhibits stable and efficient performance throughout the entire simulation period, achieving an average efficiency of 98.60% at 1 kW/m^2 and 97.70% at 0.2 kW/m^2. In contrast, the average efficiency for the INC and P&O methods under high irradiance (1 kW/m^2) with a 0.5% perturbation step size is 84.67 and 84.63%, respectively, which increases to 88.27 and 88.32% with a 0.1% step size. Under low irradiance (0.2 kW/m^2), the average tracking efficiency for INC and P&O is 98.29 and 98.28% for a 0.5% step size, and 97.52 and 97.65% for a 0.1% step size.

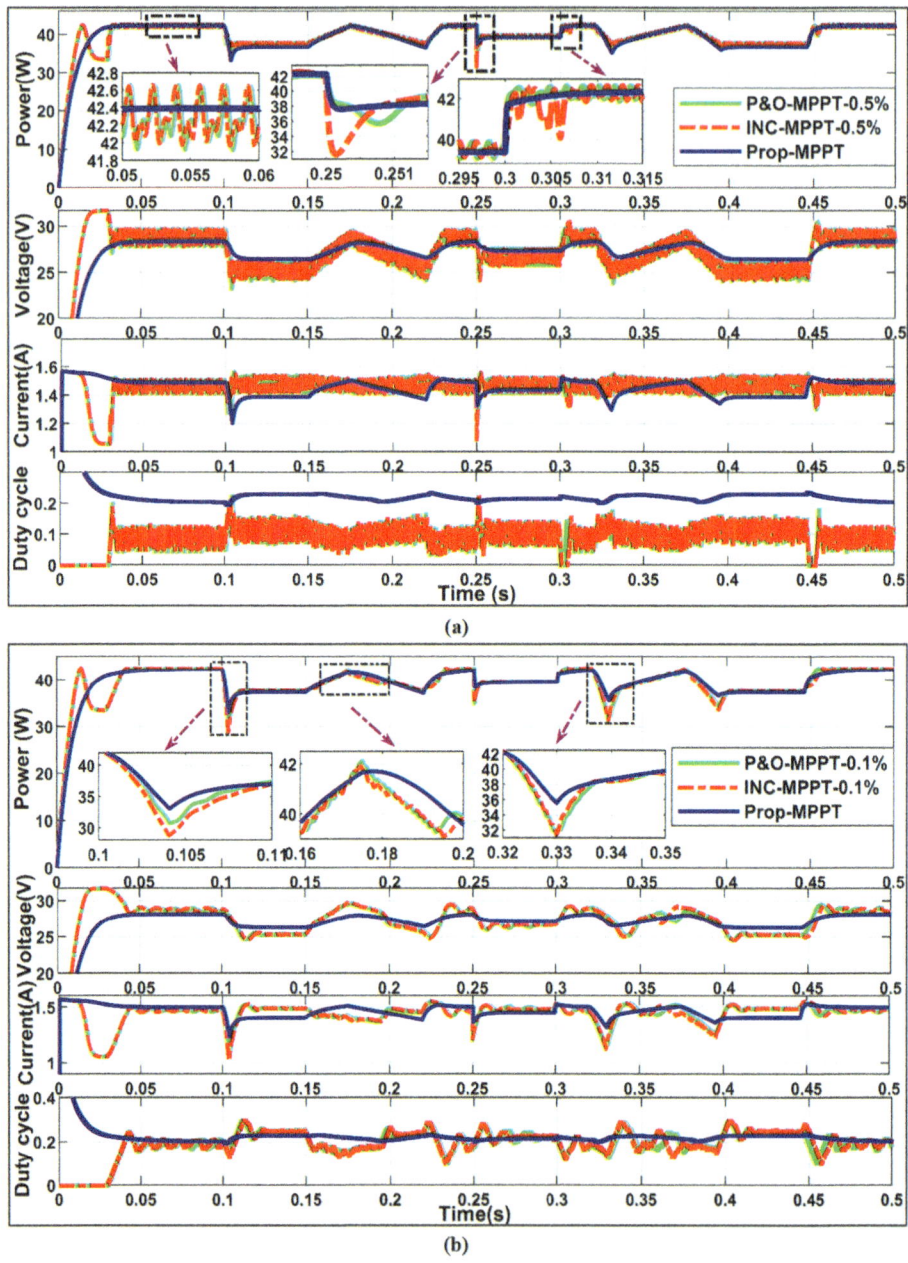

Fig. 4.53 Simulation results comparing the proposed MPPT with INC and P&O under low irradiance (0.2 kW/m^2) for perturbation step sizes of 0.5 and 0.1%

4.5 Fourth Proposed Algorithm: An Innovative MPPT Approach ...

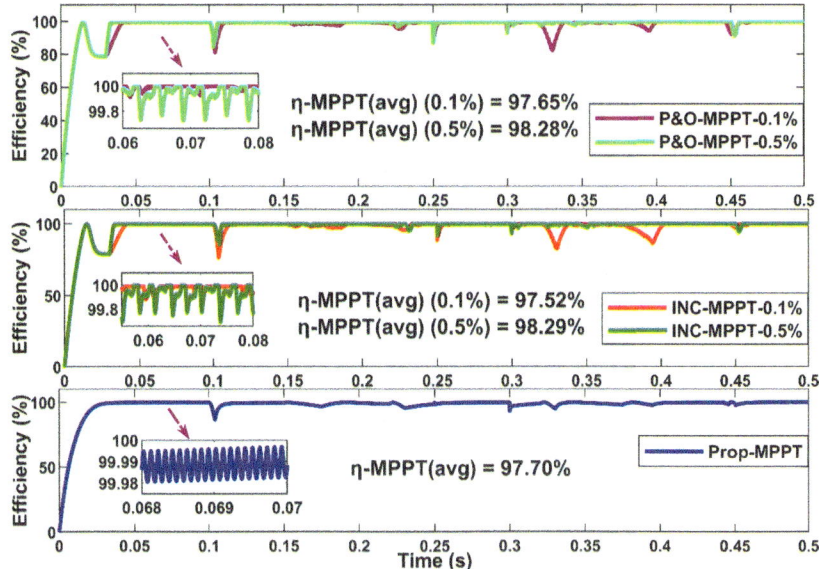

Fig. 4.54 Tracking efficiency of the MPPT methods under low irradiance and the first temperature scenario

4.5.2.2 Simulation Results According to the Second Temperature Condition Scenario Under High and Low Insolation Level

Figures 4.55, 4.56, 4.57 and 4.58 illustrate the simulation outcomes of the INC and P&O methods compared to the proposed MPPT strategy under the second temperature scenario. This scenario explores two fixed irradiance conditions: high irradiance at 1 kW/m^2 and low irradiance at 0.2 kW/m^2.

The comparative analysis of Figs. 4.55, 4.56, 4.57 and 4.58 indicates that the proposed MPPT method significantly outperforms the conventional INC and P&O approaches under both irradiance levels and temperature conditions. The proposed method consistently tracks the MPP with high precision, rapid convergence, and minimal response time, maintaining zero oscillation throughout the simulation.

As observed in Figs. 4.56 and 4.58, the proposed MPPT method achieves a dynamic tracking efficiency exceeding 99.8% under high irradiance and 99.5% under low irradiance, with average tracking efficiencies of 98.94 and 97.69%, respectively. In contrast, Figs. 4.55 and 4.57 reveal that the INC and P&O methods struggle to track the MPP, especially when the perturbation step size is set to 0.1%. This small step size limits their response speed, leading to delayed MPP attainment. When a larger step size of 0.5% is applied, both methods encounter significant oscillations around the MPP, causing further inefficiencies, as shown in Fig. 4.55b.

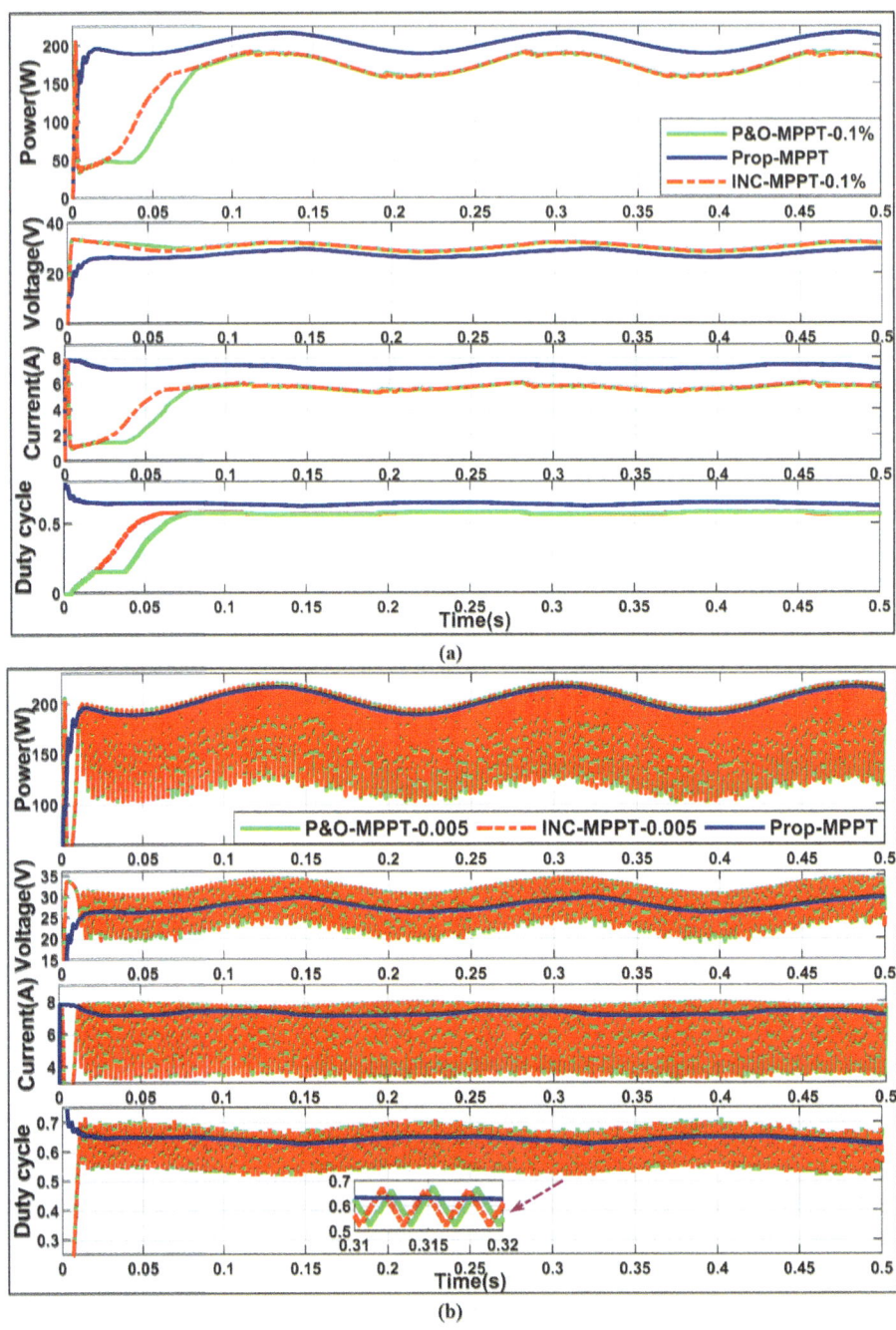

Fig. 4.55 Simulation results for the proposed MPPT method compared to INC and P&O under high irradiance (1 kW/m^2) with 0.1 and 0.5% perturbation step sizes

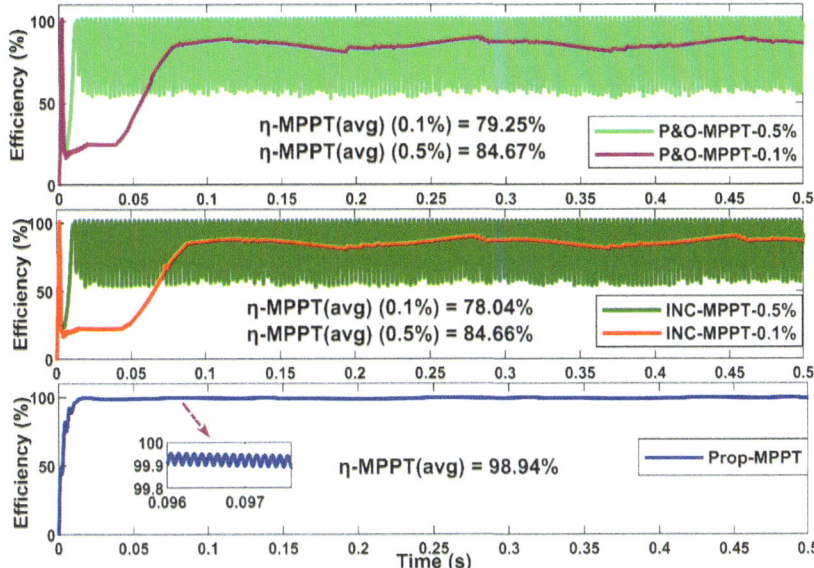

Fig. 4.56 Tracking efficiency of MPPT methods under high irradiance in the second temperature scenario

Detailed in Figs. 4.56 and 4.58, the dynamic tracking efficiency of the INC and P&O methods declines sharply with a 0.5% perturbation step size, falling below 60%, with average tracking efficiencies of 84.67 and 84.66%, respectively, under high irradiance. When using a 0.1% step size, the INC and P&O methods achieve slightly improved average efficiencies of 79.25 and 78.04%, but their dynamic tracking efficiency remains below 30%, as indicated in Fig. 4.56. Under low irradiance, Fig. 4.58 demonstrates that both INC and P&O methods yield similar tracking efficiencies, regardless of whether a 0.5% or 0.1% step size is used, reaching average efficiencies of 78.27 and 75.49%, respectively.

Table 4.7 provides a comprehensive summary of the tracking efficiencies for the INC, P&O, and proposed methods across all simulated scenarios. The data highlights that the proposed MPPT method offers significant improvements in tracking efficiency, outperforming INC and P&O by approximately 9 and 8.5%, respectively.

4.6 Summary

In this chapter, four newly developed MPPT approaches are proposed to significantly enhance the tracking performance and address the challenges commonly faced by existing MPPT techniques.

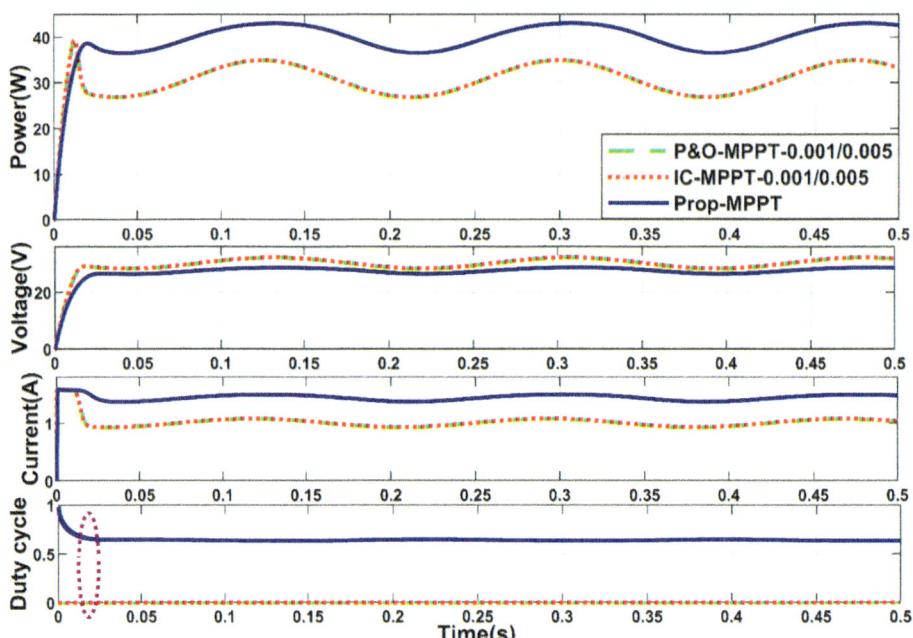

Fig. 4.57 Simulation results for the proposed MPPT method compared to INC and P&O under low irradiance (0.2 kW/m^2) with 0.1 and 0.5% perturbation step sizes

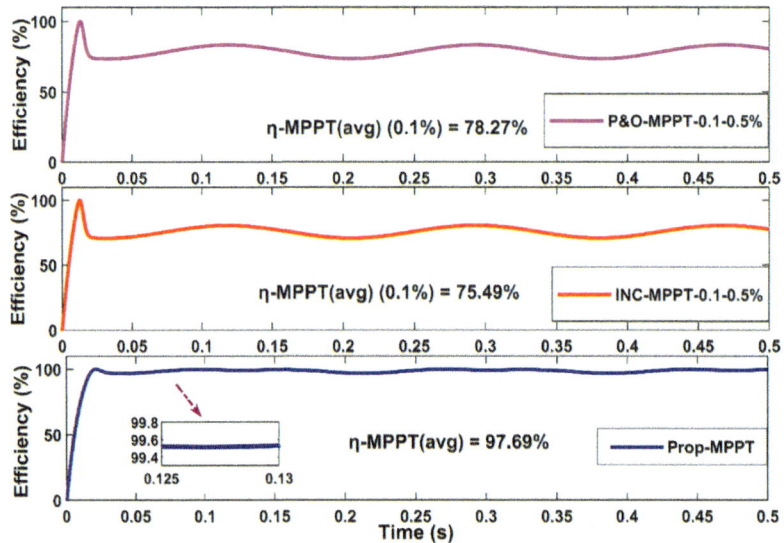

Fig. 4.58 Tracking efficiency of MPPT methods under low irradiance in the second temperature scenario

4.6 Summary

Table 4.7 Summary of tracking efficiencies for INC, P&O, and the proposed MPPT strategy under various simulation conditions

Scenario type		Minimum efficiency (%)					Maximum efficiency (%)					Average efficiency (%)				
		P&O		INC		Proposed	P&O		INC		Proposed	P&O		INC		Proposed
		Perturbation step size					Perturbation step size					Perturbation step size				
		0.5%	0.1%	0.5%	0.1%	–	0.5%	0.1%	0.5%	0.1%	–	0.5%	0.1%	0.5%	0.1%	–
First scenario	1 kW/m²	50.30	54.50	49.80	55.30	98.50	99.96	99.92	99.53	99.65	99.94	84.63	88.32	84.67	88.27	98.60
	0.2 kW/m²	87.56	82.92	88.32	77.60	88.63	99.96	99.96	99.93	99.93	99.95	98.28	97.65	98.29	97.52	97.70
Second scenario	1 kW/m²	58.30	25.60	58.80	23.60	99.80	99.63	83.20	99.60	80.40	99.95	84.67	79.25	84.66	78.04	98.94
	0.2 kW/m²	73.73	73.73	71.34	71.34	97.17	83.60	83.60	81.10	81.10	99.99	78.27	78.27	75.49	75.49	97.69

The First Proposed Approach presented focuses on adapting to temperature variations with enhanced tracking speed and minimized steady-state fluctuations at the MPP. This strategy can be integrated with a variety of existing MPPT techniques to improve their tracking accuracy, reduce response times, and minimize power losses. The proposed approach achieves fast and accurate tracking under critical temperature conditions without adding complexity to the implementation process. Simulations conducted using MATLAB/Simulink, under diverse temperature and irradiance scenarios, demonstrate the effectiveness of this approach in improving tracking efficiency and reducing power losses under both steady-state and dynamic conditions. The results show that integrating this novel approach with conventional algorithms, such as P&O, INC, and even newer methods like the modified MPP-Locus algorithm, yields a considerable improvement in average efficiency. Specifically, the average efficiency of the conventional P&O, INC, and modified MPP-Locus algorithms is reported at 98.85, 98.80, and 98.81%, respectively, whereas the improved versions achieve efficiencies of 99.18, 99.06, and 99.12%, respectively. Furthermore, the proposed approach reduces response times by 5.25, 2.06, and 2.57 ms compared to the conventional INC, P&O, and modified MPP-Locus algorithms, while reducing steady-state power ripples from 1, 2, and 0.6 W to negligible levels.

The second MPPT strategy aims to optimize photovoltaic power extraction under challenging irradiance and load conditions while maintaining negligible steady-state oscillations at the MPP. This approach was tested in both MATLAB/Simulink and Proteus environments, allowing for comprehensive performance analysis under abrupt irradiance and load variations. The Simulink environment was utilized to evaluate performance metrics such as response to sudden changes, while Proteus was used to demonstrate a cost-effective implementation using an Arduino board and LCD display. The results from these environments indicate that the proposed method significantly improves convergence speed during rapid changes in irradiance and load and eliminates oscillations at steady state. The proposed approach achieves a tracking time of less than 9.6 ms in the Simulink environment and 0.24 μs in the Proteus environment, making it approximately five to six times faster than the INC and P&O methods. Tracking efficiencies range from 99.40 to 99.76% across all simulated conditions.

The third approach, ASSTA, introduces an innovative tracking mechanism that employs the concept of the Theta angle (θ), which is defined as the arctangent of the change in PV power to the change in PV voltage (arctan(dP/dV)) and its derivative with respect to PV voltage (dθ/dV). This novel approach was tested using MATLAB/Simulink in a standalone PV system. Comparative analysis with other MPPT techniques—including INC, P&O, Variable Step Size INC (VSSINC), and PSO—demonstrates that ASSTA offers superior tracking accuracy and reduces power loss. During rapid climatic changes, ASSTA ensures fast convergence, achieving a tracking time of less than 0.017 s and negligible ripples in steady-state conditions. The average tracking efficiency of ASSTA across different scenarios was observed to be between 99.10 and 99.84%, highlighting its exceptional performance.

The final MPPT approach introduced in this chapter aims to improve tracking performance under extreme temperature variations. This novel strategy effectively addresses key limitations of conventional MPPT methods, including oscillations around the MPP, slow convergence rates, tracking loss during rapid temperature changes, and poor performance under low irradiance. Simulations in the MATLAB/Simulink environment were conducted under a range of temperature and irradiance conditions. The results indicate that this approach decreases tracking time by a factor of five, eliminates steady-state ripples around the MPP, and improves average tracking efficiency by 9.04 and 8.51% compared to INC and P&O strategies, respectively.

References

1. M. Mao, L. Cui, Q. Zhang, K. Guo, L. Zhou, H. Huang, Classification and summarization of solar photovoltaic MPPT techniques: a review based on traditional and intelligent control strategies. Energy Rep. **6** (2020). https://doi.org/10.1016/j.egyr.2020.05.013
2. M. Abdel-Salam, M.T. EL-Mohandes, M. Goda, History of maximum power point tracking, in *Green Energy and Technology* (2020). https://doi.org/10.1007/978-3-030-05578-3_1
3. X. Li, H. Wen, Y. Hu, L. Jiang, Drift-free current sensorless MPPT algorithm in photovoltaic systems. Sol. Energy **177** (2019). https://doi.org/10.1016/j.solener.2018.10.066
4. C. Abdelkhalek, E.L.B. Said, A. Younes, A. Hassan, A study and implementation of interleaved boost converter with a novel MPPT tactic for PV systems, in *2020 IEEE 2nd International Conference on Electronics, Control, Optimization and Computer Science (ICECOCS)* (2020), pp. 1–6
5. N.H. Baharudin, T.M.N.T. Mansur, F.A. Hamid, R. Ali, M.I. Misrun, Topologies of DC-DC converter in solar PV applications. Indones. J. Electr. Eng. Comput. Sci. **8**(2) (2017). https://doi.org/10.11591/ijeecs.v8.i2.pp368-374
6. E. Román, R. Alonso, P. Ibañez, S. Elorduizapatarietxe, D. Goitia, Intelligent PV module for grid-connected PV systems. IEEE Trans. Ind. Electron. **53**(4) (2006). https://doi.org/10.1109/TIE.2006.878327
7. C. Abdelkhalek, E.L.B. Said, A. Younes, Ripples amplitude minimizing of the output boost converter using an innovative MPPT controller for PV systems applications, in *2020 IEEE 2nd International Conference on Electronics, Control, Optimization and Computer Science (ICECOCS)* (2020), pp. 1–6. https://doi.org/10.1109/ICECOCS50124.2020.9314628
8. C. Abdelkhalek, E.L.B. Said, A. Younes, An improved MPPT tactic for PV system under temperature variation, in *2019 8th International Conference on Systems and Control, ICSC 2019* (2019). https://doi.org/10.1109/ICSC47195.2019.8950508
9. A. Chellakhi, S. El Beid, Y. Abouelmahjoub, An improved maximum power point approach for temperature variation in pv system applications. Int. J. Photoenergy **2021** (2021). https://doi.org/10.1155/2021/9973204
10. X. Li, H. Wen, W. Xiao, A modified MPPT technique based on the MPP-locus method for photovoltaic system, in *Proceedings IECON 2017—43rd Annual Conference of the IEEE Industrial Electronics Society* (2017). https://doi.org/10.1109/IECON.2017.8216394

11. M.G. Batarseh, M.E. Za'ter, Hybrid maximum power point tracking techniques: A comparative survey, suggested classification and uninvestigated combinations. Solar Energy **169** (2018). https://doi.org/10.1016/j.solener.2018.04.045
12. Á.A. Bayod-Rújula, J.A. Cebollero-Abián, A novel MPPT method for PV systems with irradiance measurement. Sol. Energy **109**(1) (2014). https://doi.org/10.1016/j.solener.2014.08.017
13. A. Chellakhi, S. El Beid, Y. Abouelmahjoub, Implementation of a novel MPPT tactic for PV system applications on MATLAB/simulink and proteus-based arduino board environments. Int. J. Photoenergy **2021** (2021). https://doi.org/10.1155/2021/6657627
14. C. Abdelkhalek, E.B. Said, A. Younes, A novel MPPT tactic for PV systems with fast-converging speed and zero oscillation, in *2020 5th International Conference on Renewable Energies for Developing Countries, REDEC 2020* (2020). https://doi.org/10.1109/REDEC49234.2020.9163606
15. A. Chellakhi, S. El Beid, High-efficiency MPPT strategy for PV systems: ripple-free precision with comprehensive simulation and experimental validation. Results Eng. 103230 (2024). https://doi.org/10.1016/J.RINENG.2024.103230
16. A. Chellakhi, S. El Beid, Y. Abouelmahjoub, A novel theta MPPT approach based on adjustable step size for photovoltaic system applications under various atmospheric conditions. Energy Syst. (2022). https://doi.org/10.1007/s12667-022-00519-2
17. A. Chellakhi, S. El Beid, Y. Abouelmahjoub, An innovative fast-converging speed MPPT approach without oscillation for temperature varying in photovoltaic systems applications. Energy Sour. Part A Recover. Util. Environ. Eff. https://doi.org/10.1080/15567036.2022.2058121 (2022)

Experimental Validation 5

5.1 Introduction

Isolated photovoltaic systems play a vital role in numerous applications, facilitating the distribution of electricity worldwide. In this chapter, we present the implementation of the practical test bench for the proposed standalone PV system and its components, conducted at the CISIEV Team Laboratory, Faculty of Sciences and Technology, Cadi Ayad University, Marrakech, Morocco. The chapter is organized into two main sections: the first section describes the experimental setup prototype of the standalone PV system, while the second section presents the experimental validation results and discusses the performance of the four newly proposed MPPT approaches.

5.2 Experimental Setup Prototype of the Used Standalone PV System

To evaluate the performance of the proposed novel MPPT methods under practical conditions, an experimental prototype of a standalone PV system was developed, as depicted in Fig. 5.1. The proposed off-grid PV system consists of a PV array that supplies DC power through a DC-DC boost converter, while MPPT algorithm implementation is handled using a dSPACE DS1104 platform interface. Current and voltage are measured using LA 25-NP and MX 9030 sensors, respectively, while solar irradiance and temperature are measured using a Kipp & Zonen pyranometer (sensitivity: 14.69×10^{-3} mV) and a PN junction silicon diode, respectively. The following subsections provide a detailed overview of the main components of the PV system prototype.

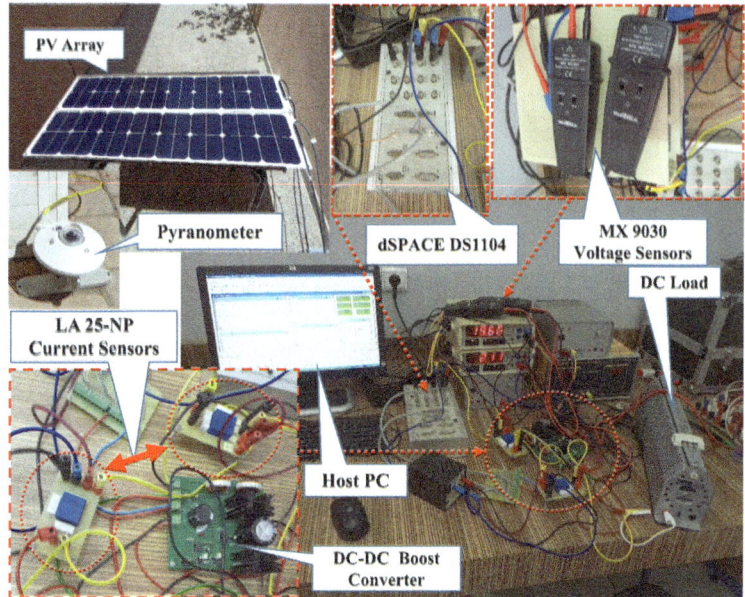

Fig. 5.1 Experimental prototype setup of the proposed standalone PV system developed in the CISIEV Team Laboratory

5.2.1 PV Array

The PV array used in this setup is constructed from two PV panels connected in parallel, as shown in Fig. 5.1. Table 5.1 presents the parameters of the overall system including those of the PV panel under STC, as reported by the manufacturer. However, it should be noted that the PV panel's performance tends to degrade over time, which means the actual power output may be lower than the datasheet specifications. To account for this degradation, preliminary experimental tests were conducted to evaluate the performance of the PV panel under real-world conditions before integrating it into the PV system. The resulting P-V and I-V curves for the PV array are shown in Fig. 5.2, depicting data collected over two days (August 11 and 12, 2022) in Marrakech. The corresponding insolation and ambient temperature profiles are shown in Fig. 5.3.

Figure 5.4 illustrates the conventional method for tracing PV characteristics, which involves using a variable resistor to generate P-V and I-V curves. Although straightforward, this method is inefficient due to the time required to incrementally adjust the resistor while recording the current and voltage values. In contrast, our approach utilizes the capabilities of the dSPACE Real-Time Interface (RTI) to streamline this process using the developed PV system. The method, designed within the Simulink environment, is

5.2 Experimental Setup Prototype of the Used Standalone PV System

Table 5.1 Parameters used in the practical test bench of the proposed off-grid PV system

Parameters	Designation	Specifications
PV panel		
MPP	P_{MPP}	60 W
Output power tolerance		±5
MPP voltage	V_{MPP}	17.6 V
MPP current	I_{MPP}	3.4 A
Open circuit voltage	V_{OC}	21.2 V
Short circuit current	I_{SC}	4.2 A
DC-DC converter	Boost	
Input capacitance	C_{IN}	1000×10^{-6} F, 100 V
Output capacitance	C_{OUT}	470×10^{-6} F, 400 V
Inductance	L	165×10^{-6} H
MOSFET	P30N60E	[1]
MOSFET driver	MCP 1406	[2]
Switching frequency	f	30 kHz
Sensors		
Voltage sensor	MX 9030	[3]
Current sensor	LA 25-NP	[4]
Solar sensor	Kipp & Zonen pyranometer	[5]
Temperature sensor	PN junction silicon diode	[6]

compiled for use in the dSPACE ControlDesk platform. It controls the gate of the DC-DC boost converter by dynamically adjusting the duty ratio from 0.1 to 0.85, allowing the complete PV curve to be traced efficiently. The resulting P-V and I-V characteristics are shown in Fig. 5.5, providing real-time insights into PV behavior.

5.2.2 DC-DC Boost Converter

The main components of the designed DC-DC boost converter are shown in Fig. 5.6. These include capacitors, inductors, a diode, voltage regulators, a MOSFET, and a MOSFET driver, along with other elements such as a fuse for protection and heat dissipation. Detailed specifications of these components are provided in Table 5.1.

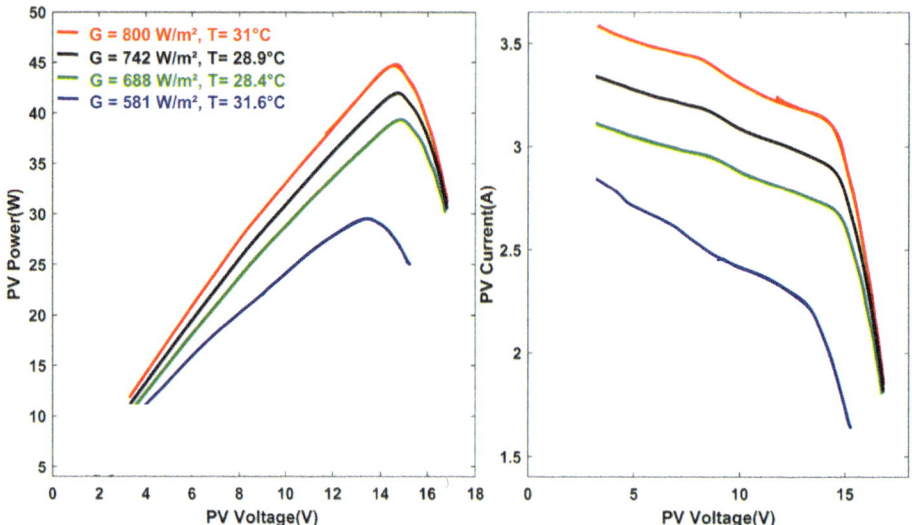

Fig. 5.2 Experimental results of the P-V and I-V characteristics of the PV array used

5.2.3 DSPACE DS1104 Controller Card

The dSPACE system, based on the DS1104 R&D controller board, is an advanced prototyping system developed by the German company "dSPACE GmbH". It is specifically designed for high-speed multivariable digital controller development and real-time simulation across various fields. The DS1104 R&D controller board is especially advantageous due to its cost-effectiveness for Rapid Control Prototyping (RCP) applications, including robotics, electrical machines, and the aerospace industry.

The dSPACE DS1104 system is comprised of two main components: hardware and software. The hardware components include the DS1104 R&D controller card and the CP1104 connector panel, connected via a master I/O ribbon cable, as illustrated in Fig. 5.7. The CP1104 connector panel serves as the interface between the DS1104 controller card and the system devices.

The CP1104 connector panel features numerous interface connectors, as shown in Fig. 5.7. In our work, two key connectors were utilized. The first is the slave I/O PWM connector, featuring 37 pins, one of which is used to produce the required PWM signal for the DC-DC converter. The second set of connectors are the BNC connectors, which provide a voltage range of ± 10 V and are split into two groups: Analog-to-Digital Converters (ADC) (CP1 to CP8) and Digital-to-Analog Converters (DAC) (CP9 to CP16).

5.2 Experimental Setup Prototype of the Used Standalone PV System

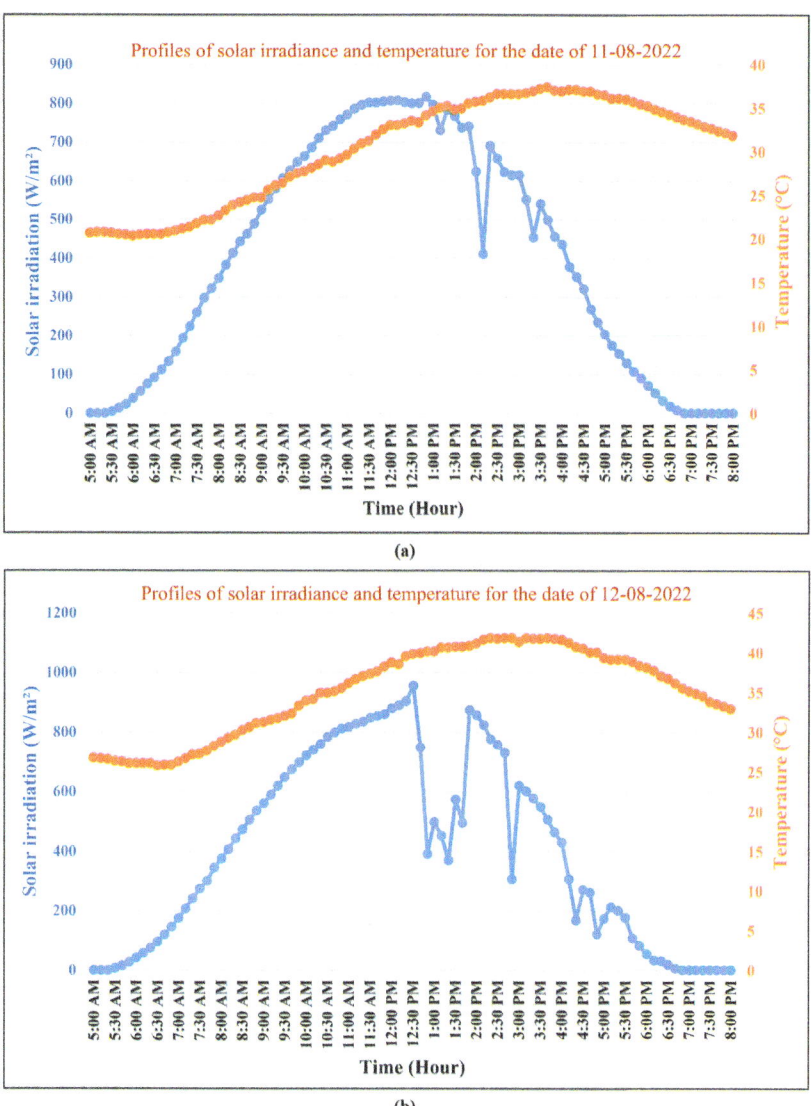

Fig. 5.3 Real solar irradiance and ambient temperature conditions for two days in the city of Marrakech; **a** 11 and **b** 12 August, 2022

The software component of the dSPACE system includes the dSPACE ControlDesk program, depicted in Fig. 5.8, which serves as the graphical user interface enabling real-time communication between Simulink models and physical hardware. The ControlDesk software performs several crucial tasks, including:

156   5 Experimental Validation

Fig. 5.4 Conventional method for plotting the I-V characteristics of the PV array

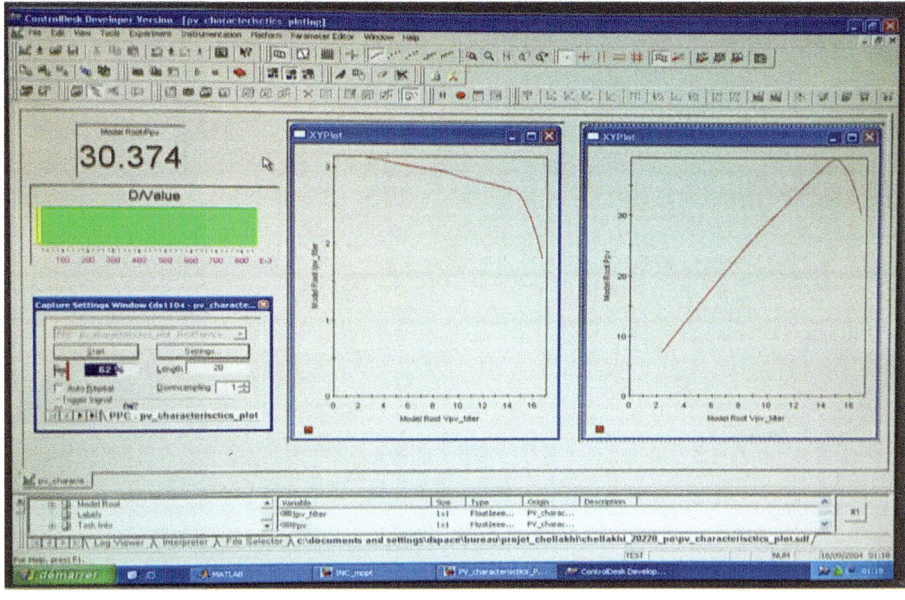

Fig. 5.5 A screenshot of the dSPACE ControlDesk environment constructed for plotting P-V and I-V characteristics

- Providing an interface for downloading controller models to the DS1104 R&D controller board.
- Interfacing with the entire system's inputs.
- Managing and supervising the DS1104 controller board.

5.2 Experimental Setup Prototype of the Used Standalone PV System

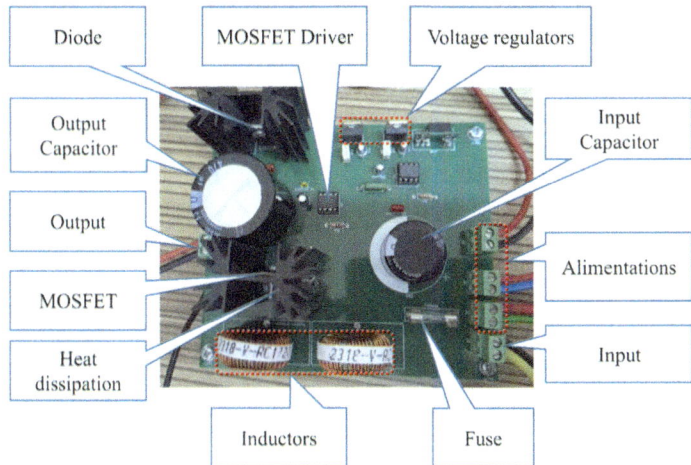

Fig. 5.6 Illustration of the principal components of the DC-DC boost converter used in the practical validation

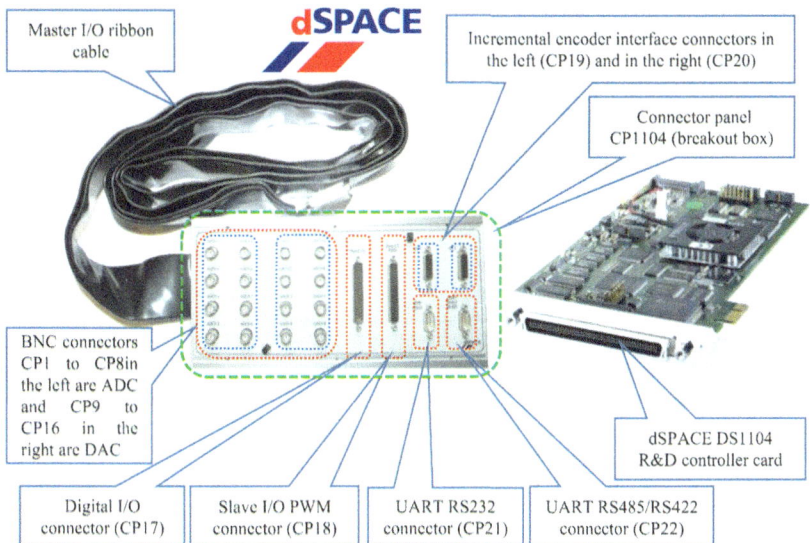

Fig. 5.7 Illustration of the hardware components of the DS1104 R&D controller board

- Monitoring, displaying, calibrating, and plotting any calculated variable in real-time on the PC display.
- Supporting Rapid Control Prototyping (RCP).
- Enabling Hardware-In-the-Loop (HIL) simulation.

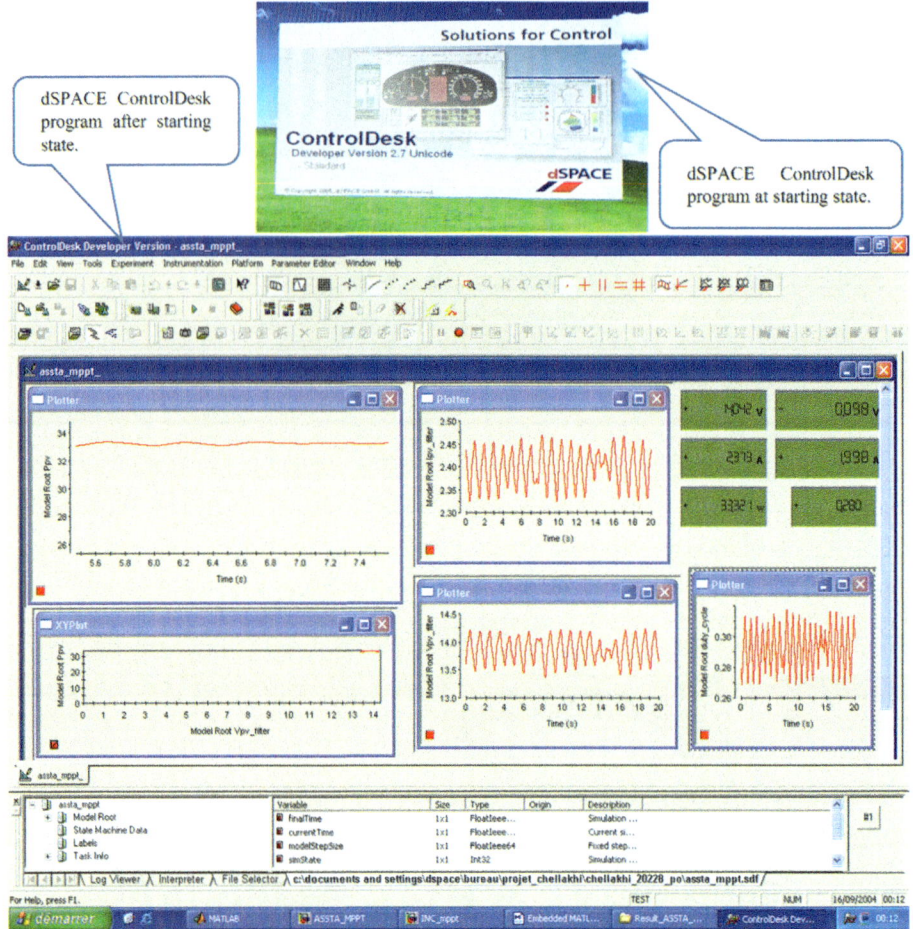

Fig. 5.8 Illustration of the interface of the dSPACE ControlDesk program

5.2.4 Working Procedure of the Used Standalone PV System

Figure 5.9 presents a schematic diagram illustrating the connections between the components of the experimental prototype of the proposed photovoltaic system. Initially, a Simulink model of the PV system is created within the Simulink environment, aimed at implementing the MPPT algorithm using the dSPACE system, as shown in Fig. 5.10. The model consists of four principal blocks adjacent to the MPPT block:

1. **ADC Block for Digital Signal Processing (DSP)**: Converts the analog values of the measured PV voltage and current into numerical values required by the MPPT block.

5.2 Experimental Setup Prototype of the Used Standalone PV System 159

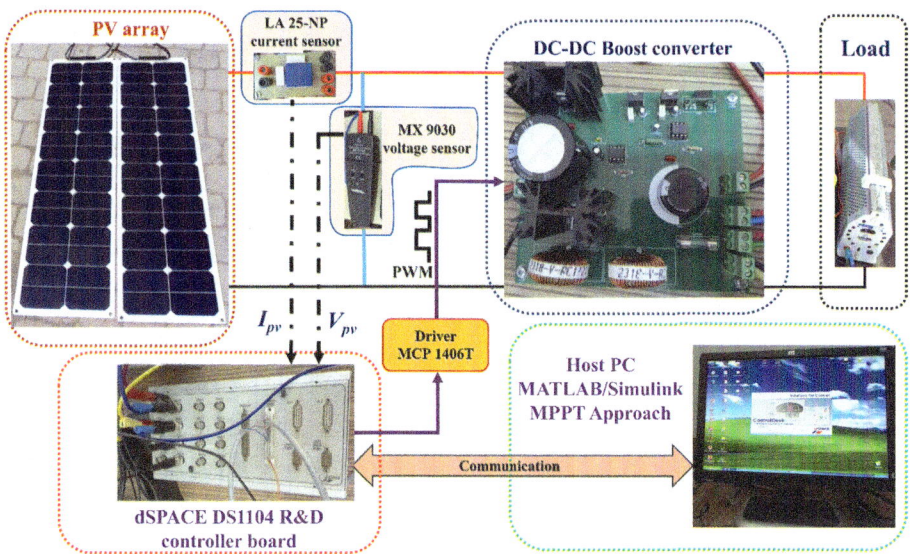

Fig. 5.9 Schematic diagram explaining the connections between the components of the experimental prototype of the proposed photovoltaic system

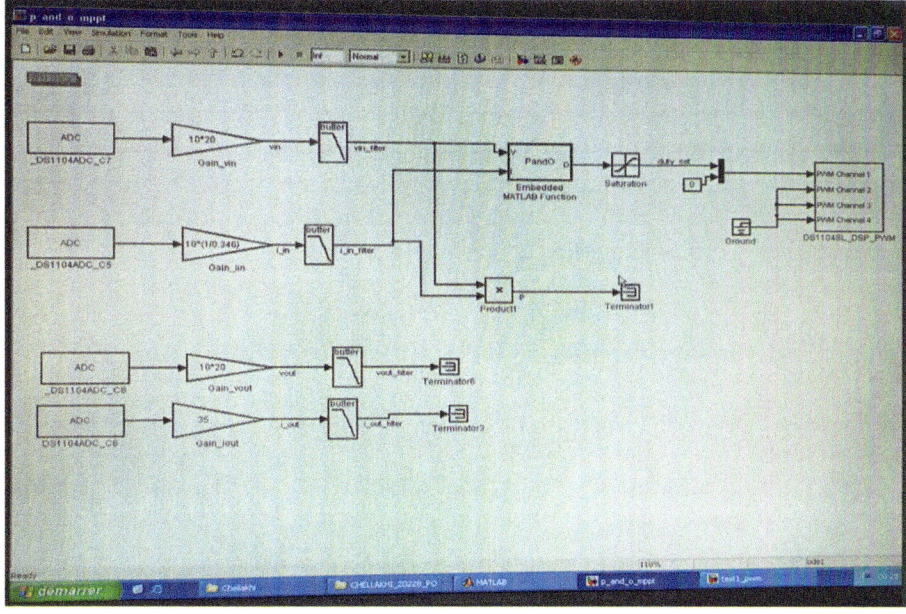

Fig. 5.10 Screenshot of the implementation diagram of the MPPT algorithm in the Simulink environment using the dSPACE interface

2. **Analog Filter Block**: Used to eliminate high-frequency noise signals.
3. **Calibration Block**: Used to ensure accuracy in measurements.
4. **DS1104SL-DSP-PWM Block**: Responsible for generating the PWM signal needed for controlling the DC-DC boost converter.

After the Simulink model is constructed, it is compiled to generate a system description file (SDF), which is then downloaded to the DS1104 board. The DS1104 board performs the MPPT approach, while dSPACE ControlDesk is used to monitor and display the variables of the Simulink model in real-tim.

5.3 Experimental Validation Results of the Four Novel Proposed MPPT Approaches

The experimental validation of the four-novel proposed MPPT approaches was successfully conducted using the test bench of the proposed PV system shown in Figs. 5.1 and 5.9. This validation was performed in the laboratory of the CISIEV Team at the Faculty of Sciences and Technology (FST) of Cadi Ayyad University in Marrakech during August 2022. The practical validation experience formed an essential part of this research, offering insights into how real-world issues can be effectively addressed in a photovoltaic system.

Two major challenges were encountered during the experiments. First, the unstable weather conditions made it difficult to repeat the same experiment for multiple algorithms under consistent environmental parameters. Second, the lack of specialized components hindered our ability to induce rapid temperature or irradiance variations. Despite these challenges, the experimental results were invaluable in identifying and solving practical issues, ultimately demonstrating the robustness and effectiveness of the proposed MPPT approaches under real conditions.

5.3.1 Experimental Results of the First Proposed MPPT Approach

The first developed MPPT approach, specifically designed to address the challenges of MPP tracking in conventional methods under varying temperature conditions, was successfully validated in practice by integrating it with traditional algorithms such as the P&O and INC MPPT algorithms. Given the slow variation rate of ambient temperature, as shown in Fig. 5.3, and the inability to rapidly change temperature, the validation considered only the stable temperature conditions.

The practical results of the improved P&O (IMP-P&O) and improved INC (IMP-INC) approaches, compared to the traditional P&O and INC methods, respectively, are depicted

5.3 Experimental Validation Results of the Four Novel Proposed ...

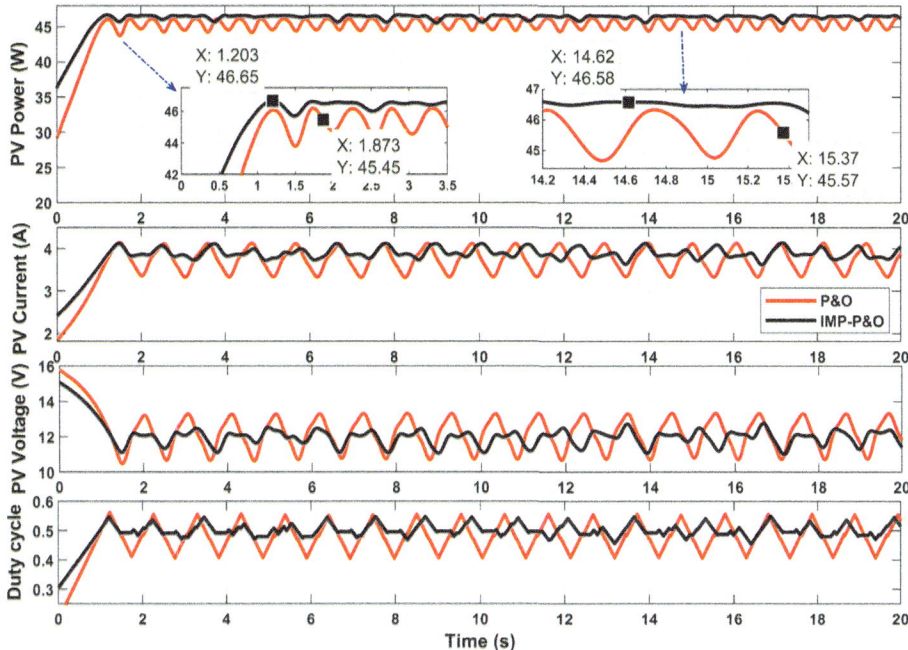

Fig. 5.11 Experimental results of the IMP-P&O MPPT compared to the traditional P&O method

in Figs. 5.11 and 5.12. The validation test was conducted on August 8, 2022, under insolation and temperature conditions of approximately 820 W/m² and 31 °C, respectively.

The experimental results illustrated in Figs. 5.11 and 5.12 show the waveforms of PV power, current, voltage, and duty cycle. It is evident that the improved MPPT methods outperformed the conventional algorithms. Specifically, the improved approaches significantly enhanced tracking speed, reduced oscillations around the MPP, and maximized power extraction, as indicated by the power curves.

5.3.2 Experimental Results of the Second Proposed MPPT Approach

The experimental validation of the second proposed MPPT approach was carried out under two types of weather conditions: stable and rapidly changing insolation. To emulate a fast-changing irradiance scenario, a prototype was developed as depicted in Fig. 5.13. This setup utilized two identical PV panels (the PV array shown in Fig. 5.9) in parallel, with a switch in series with one of the panels. Initially, the switch was closed, allowing the system to operate at the MPP of both panels. Subsequently, the switch was opened to emulate a rapid decrease in insolation, though this operation was only performed after

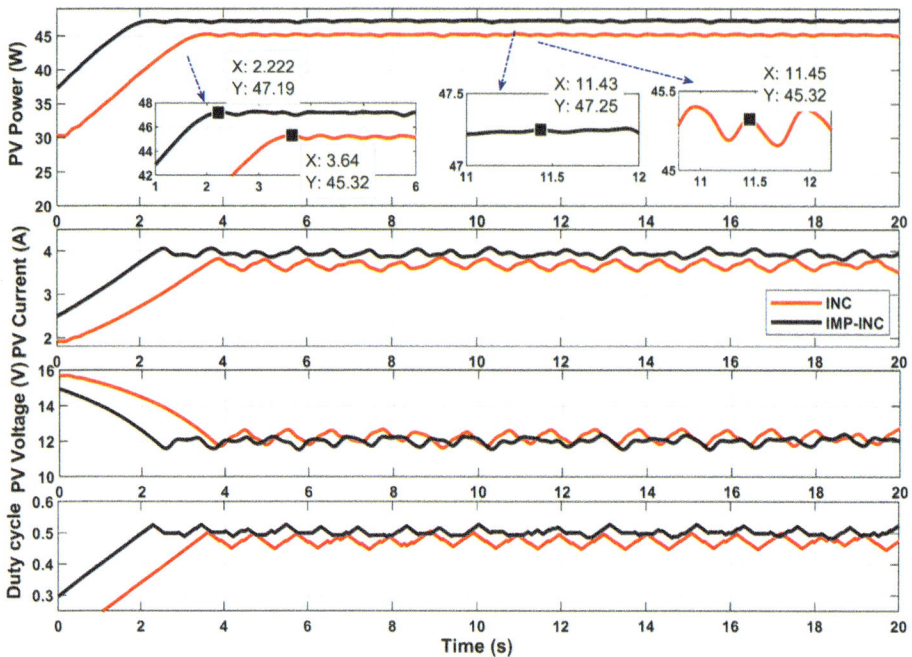

Fig. 5.12 Experimental results of the IMP-INC MPPT compared to the traditional INC method

the system reached a steady state. Similarly, closing the switch emulated a rapid increase in insolation.

The experimental results of the novel MPPT approach, compared to the P&O and INC techniques under stable and rapid climatic conditions, are shown in Figs. 5.14 and 5.15, respectively. These results were obtained under stable conditions on August 12, 2022, with solar irradiance and temperature of approximately 930 W/m^2 and 36 °C. In the second scenario, conducted on August 14, 2022, solar irradiation varied abruptly between 830 and 415 W/m^2, with a fixed temperature of 30.2 °C.

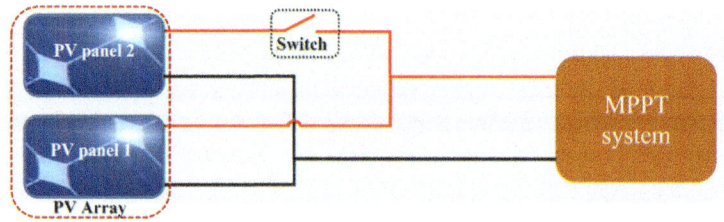

Fig. 5.13 Architecture of the used PV system to emulate a fast insolation change situation

5.3 Experimental Validation Results of the Four Novel Proposed ...

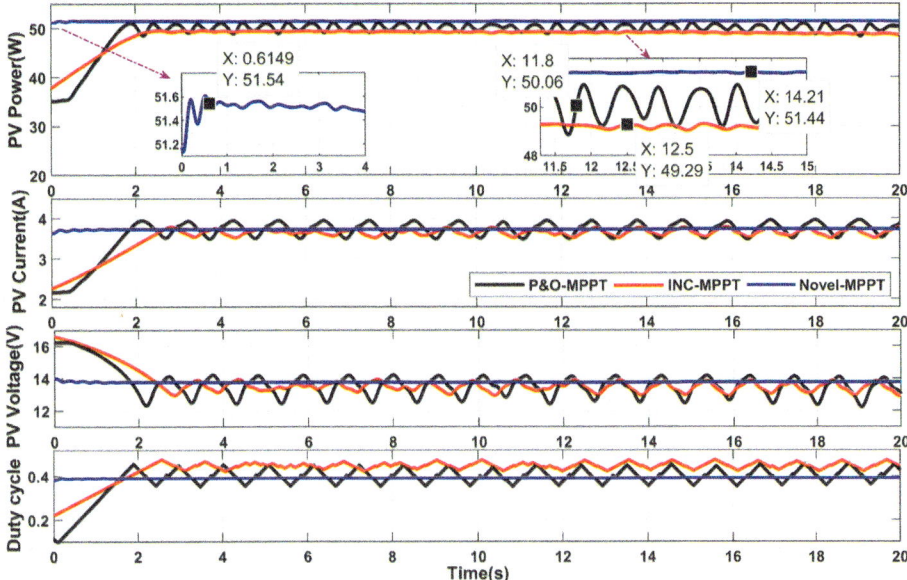

Fig. 5.14 Experimental results of the novel MPPT approach compared to INC and P&O techniques under normal conditions

Under stable conditions, Fig. 5.14 illustrates that the novel proposed MPPT method provided faster tracking velocity (less than 0.62 s) and robust tracking of the expected MPP with zero steady-state fluctuation compared to the P&O and INC methods, which exhibited noticeable fluctuations, slower tracking, and power losses.

Figure 5.15 presents the tracking performance under rapid insolation changes. The power curves indicate that traditional techniques showed poor tracking compared to the novel approach. Specifically, the INC method lost its tracking ability after rapid irradiance changes due to the small step size used in duty cycle perturbation, as seen in its duty cycle curves (Fig. 5.15a). The P&O algorithm, meanwhile, showed large oscillations and high-power loss due to poor MPP tracking under both steady-state and rapid conditions. In contrast, the novel developed MPPT approach demonstrated robustness during both steady-state and abrupt insolation variations, achieving faster tracking, higher speed convergence, and improved tracking performance.

5.3.3 Experimental Results of the Third Proposed MPPT Approach

The practical validation of the third newly proposed MPPT approach, named the Adjustable Step Size Theta Approach (ASSTA), was carried out using the same prototype and under the same scenarios as the other proposed MPPT algorithms. In the first

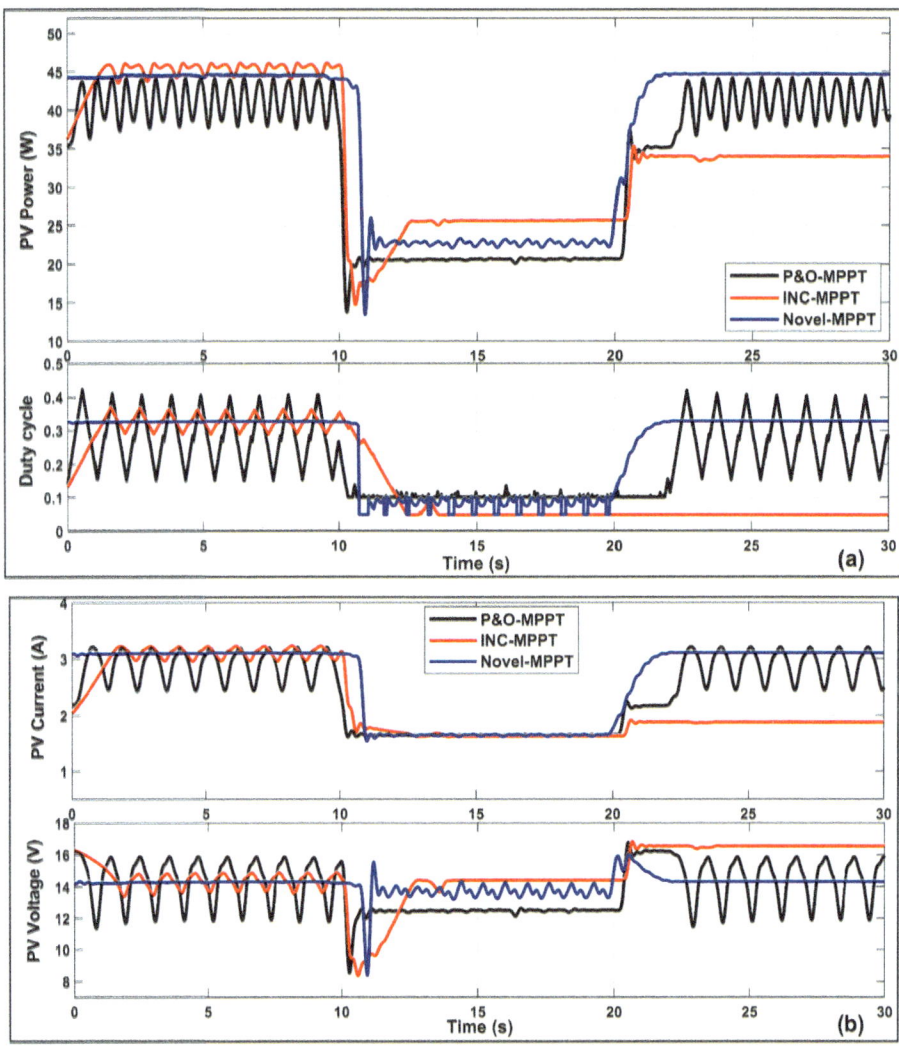

Fig. 5.15 Experimental results of the novel MPPT approach compared to INC and P&O techniques under rapid insolation variations, including **a** PV power and duty cycle waveforms and **b** PV current and voltage waveforms

scenario involving stable climatic conditions, the insolation and ambient temperature were 635 W/m^2 and 26.8 °C, respectively, on August 9, 2022. In the second scenario involving rapid insolation change, the insolation varied abruptly between 794 W/m^2 and 397 W/m^2 under a constant temperature of 33 °C on August 14, 2022.

The experimental results of the ASSTA approach, compared to the INC and P&O methods under both scenarios, are presented in Figs. 5.16 and 5.17. Under stable conditions,

5.3 Experimental Validation Results of the Four Novel Proposed ...

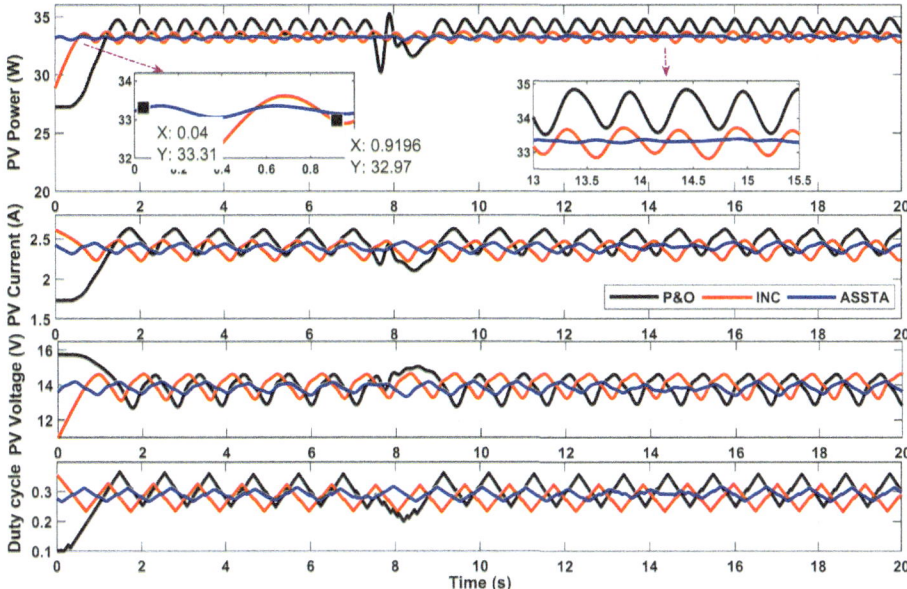

Fig. 5.16 Experimental results of the ASSTA MPPT approach compared to INC and P&O techniques under normal conditions

Fig. 5.16 shows that the ASSTA method demonstrated superior tracking performance by effectively reducing oscillations and improving tracking speed, as evident from the PV power curves. Under rapid insolation changes, Fig. 5.17 reveals that the P&O and INC algorithms struggled to track the available MPP, particularly when irradiance changed abruptly. Specifically, the P&O method exhibited large oscillations and a delayed response time, resulting in significant power losses. The INC method also lost tracking direction when insolation decreased, as seen in the power curves of Fig. 5.17a.

In contrast, the ASSTA approach maintained robust tracking of the expected MPP despite significant insolation changes. The ASSTA method tracked the MPP with zero fluctuations and high convergence speed, thereby improving dynamic tracking efficiency and maximizing energy harvested, demonstrating good robustness under dynamic conditions. The time delay between the tracking curves for the ASSTA approach and other conventional methods is due to the differing times at which the fast change in insolation was applied: ASSTA (4–10.2 s), P&O (5–11 s), and INC (5.2–11.2 s), as clearly illustrated in Fig. 5.17.

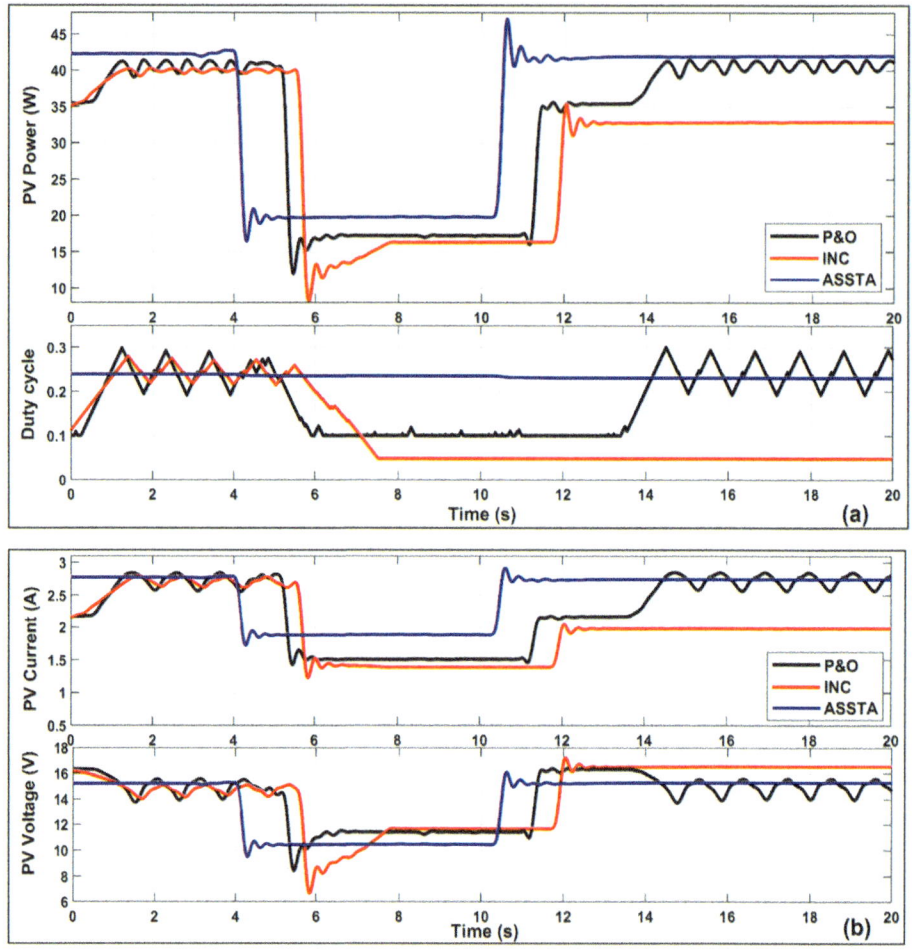

Fig. 5.17 Experimental results of the ASSTA approach compared to INC and P&O techniques under rapid insolation variation, including **a** PV power and duty cycle waveforms and **b** PV current and voltage waveforms

5.3.4 Experimental Results of the Fourth Proposed MPPT Approach

The fourth proposed MPPT approach was experimentally validated under normal climatic conditions, similar to the validation of the first developed MPPT approach, due to the challenges of replicating rapid changes in temperature or irradiance. To evaluate the performance and robustness of this new MPPT approach, it was compared under the same test conditions to the INC and P&O MPPT techniques. The experiment was conducted on August 8, 2022, with insolation and temperature conditions of 860 W/m^2 and 30.8 °C, respectively.

5.4 Summary

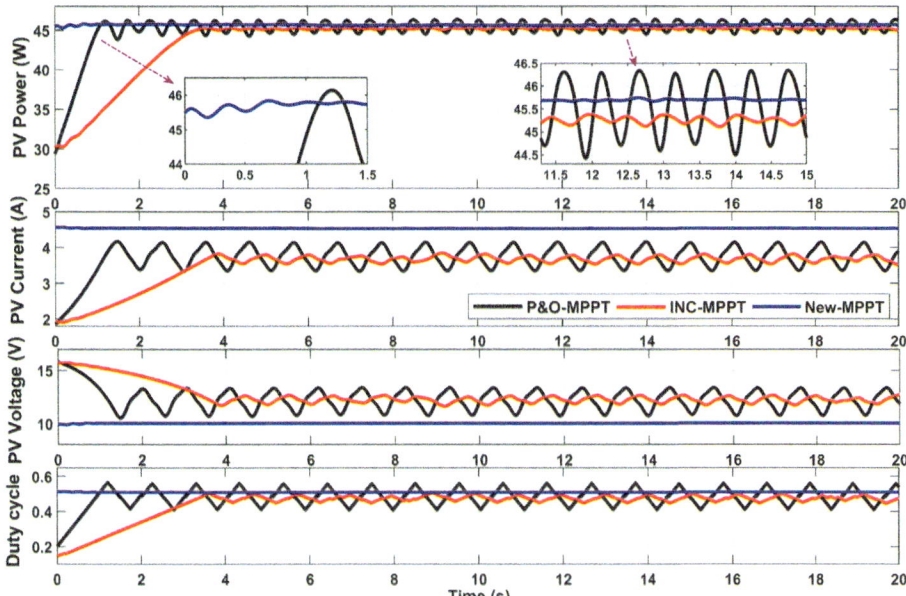

Fig. 5.18 Experimental results of the fourth new MPPT approach compared to INC and P&O algorithms under stable conditions

The experimental results obtained during this validation are shown in Fig. 5.18, which presents the waveforms of PV power, current, voltage, and the duty ratio of the boost converter. From these results, it is evident that the new proposed approach significantly outperformed the conventional methods by ensuring faster tracking time, reducing oscillation around the expected MPP, and maintaining stability, thereby enhancing the overall tracking efficiency of the PV system.

5.4 Summary

This chapter presents the practical validation of four developed MPPT approaches using the dSPACE DS1104 controller board integrated into a complete standalone photovoltaic (PV) system. Each MPPT method was successfully implemented and tested on an experimental bench. The results confirm that these novel approaches offer enhanced tracking performance compared to traditional MPPT methods, even under both normal and extreme climatic conditions. Specifically, the new approaches demonstrated faster tracking times, high convergence speeds, negligible oscillations around the MPP, and reduced power losses, thereby achieving higher tracking efficiencies. Furthermore, the experimental results are consistent with those obtained from simulations, substantiating the feasibility and simplicity of implementing the proposed MPPT approaches in real-world PV applications.

References

1. VISHAY, P30N60E MOSFET data sheet [Online], https://datasheetspdf.com/pdf/1416985/Vishay/SiHP30N60E/1. Accessed 13 Jan 2024
2. M. Technology, MCP 1406 driver data sheet [Online], http://ww1.microchip.com/downloads/en/devicedoc/20002019c.pdf. Accessed 13 Jan 2024
3. C.A. Metrix, MX 9030 sensor data sheet [Online], https://catalog.chauvin-arnoux.co.uk/uk_en/nf-gb-mx9030-mx9030.html. Accessed 13 Jan 2024
4. L. Sensors, LA 25-NP sensor data sheet [Online]. https://www.lem.com/sites/default/files/products_datasheets/la25-np.pdf. Accessed 13 Jan 2024
5. K. Zonen, Kipp & Zonen pyranometer data sheet [Online], https://www.kippzonen.com/Download/70/Brochure-Pyranometers. Accessed 13 Jan 2024
6. Meteo-shopping, PN junction silicon diode sensor data sheet [Online]. https://www.meteo-shopping.com/en/sensors/72-sonde-temperature-inox-avec-connecteur-rj.html. Accessed 13 Jan 2024

6 Conclusion and Future Research

6.1 General Conclusion

As we contemplate future energy solutions, it is crucial to power remote sites cost-effectively while minimizing greenhouse gas emissions. In this context, this book presents modeling and optimization strategies for standalone solar PV systems, aiming to supply isolated locations with safe, clean, renewable, and sustainable electricity. The primary focus of this work is to enhance the overall performance and efficiency of standalone solar PV systems by employing effective and robust MPPT approaches. These approaches ensure the optimal transfer of maximum available power to the load, especially during significant variations in weather conditions, such as abrupt changes in solar irradiation and temperature.

Chapter 1 introduces the growing trend of renewable energy adoption in Morocco, particularly solar energy. Driven by the urgent need to reduce reliance on fossil fuels and mitigate greenhouse gas emissions, Morocco's strategic geographical position and favorable climate make it an ideal candidate for renewable energy expansion. The country has committed to achieving 52% renewable energy capacity by 2030. With abundant solar resources and significant sunshine hours, Morocco has positioned itself as a global leader in solar energy. Ambitious initiatives like the Noor Ouarzazate Complex and various distributed solar power projects have significantly contributed to its energy independence and sustainable development goals. Moreover, Morocco's commitment extends to green hydrogen production, making it a key player in the global clean energy landscape. Through international partnerships and technological advancements, the country continues to drive its renewable energy transition and inspire others to embrace sustainable solutions.

This chapter also emphasizes the importance of PV technology in Morocco's renewable energy strategy, highlighting benefits such as low maintenance, reliability, and decreasing costs. However, it acknowledges challenges like PV panel inefficiency—with conversion rates below 25% influenced by environmental conditions. To address these challenges, advanced MPPT mechanisms are necessary to optimize efficiency.

Chapter 2 provides an overview of solar photovoltaic power and its applications, as well as the different types of solar PV systems. It reviews and discusses the most commonly used types of DC-DC converters. An extensive review of prevalent MPPT approaches is presented, classifying them into three categories: conventional, artificial intelligence (AI)-based, and hybrid MPPT algorithms, along with a discussion of their benefits and drawbacks. Finally, an overview of different battery technologies is provided, highlighting those most widely used in photovoltaic systems.

Chapter 3 investigates the modeling, implementation, and simulation of a complete standalone solar PV system using MATLAB/Simulink and Proteus tools. The chapter begins with an introduction to solar photovoltaic energy, followed by comprehensive modeling of the photovoltaic module. It examines the impact of environmental and electrical parameters on its characteristics and models various DC-DC converters commonly used in photovoltaic systems, such as boost, buck, interleaved boost, and three-level boost converters, within the Simulink environment. Lastly, the overall standalone photovoltaic system is simulated and implemented using conventional techniques like INC and P&O MPPT schemes under MATLAB/Simulink and Proteus software.

Chapter 4 introduces and investigates four newly proposed MPPT approaches aimed at improving tracking performance and overcoming issues encountered by existing conventional MPPT strategies. This chapter outlines the principles and validates the proposed MPPT strategies through simulation. The four proposed MPPT strategies were successfully implemented and validated through simulation, with results demonstrating the tracking accuracy of the novel MPPT techniques compared to traditional methods.

Chapter 5 focuses on the experimental prototype setup of a standalone PV system and presents the practical results along with discussions of the four newly proposed MPPT approaches. The four proposed MPPT strategies were successfully implemented and validated under practical conditions, demonstrating their applicability and performance in real-world applications.

This chapter synthesizes the findings of this work, offering a comprehensive general conclusion and outlining future directions for research and development in standalone solar PV systems. The general conclusion reiterates the significant contributions of this book to enhancing the efficiency and performance of standalone PV systems through advanced MPPT strategies. These strategies have proven effective in optimizing power transfer under varying environmental conditions, thereby supporting the viability of renewable energy solutions for remote and isolated areas.

In summary, this work significantly contributes to the field of standalone solar PV systems by addressing efficiency challenges through advanced MPPT strategies. The proposed approaches enhance the performance of PV systems, particularly under variable weather conditions, thereby promoting the viability of renewable energy solutions for remote and isolated areas. The advancements presented in this book not only support Morocco's renewable energy goals but also offer valuable insights for global efforts toward sustainable energy development.

6.2 Future Research Directions

While this book has significantly advanced the performance of standalone PV systems through the development of new MPPT approaches, the journey toward optimal energy harvesting is far from complete. The field of solar PV energy optimization is continually evolving, especially in relation to MPPT algorithms. Several promising avenues for future research emerge from this work:

- **Integration of PI/PID Controllers for Voltage or Current Control**: The MPPT approaches presented here are based on duty cycle control. Incorporating PV voltage or current control using Proportional-Integral (PI) or Proportional-Integral-Derivative (PID) controllers could substantially enhance the precision of MPP tracking. This integration may improve overall system performance by providing more accurate and responsive control mechanisms.
- **Adaptive Control Parameters Using Metaheuristic and AI Techniques**: Fixed-parameter PI or PID controllers may exhibit limited performance under varying environmental conditions. Employing adaptive control parameters optimized through metaheuristic algorithms like Particle Swarm Optimization (PSO) or Genetic Algorithms (GA), as well as artificial intelligence techniques such as Fuzzy Logic Controllers (FLC) and Artificial Neural Networks (ANN), represents a novel area for MPPT optimization. These methods could offer improved adaptability and robustness in fluctuating conditions.
- **Development of Advanced Approaches for Partial Shading Conditions**: Partial shading conditions (PSC) pose significant challenges by creating multiple local maxima in the power-voltage curve of PV systems. Proposing and developing new MPPT strategies capable of effectively handling complex PSC scenarios is essential. Such approaches would ensure optimal energy harvesting even under non-uniform irradiance, enhancing the reliability of PV systems.
- **Exploration with Various DC-DC Converter Topologies**: This work primarily employed DC-DC boost converters for implementing the proposed MPPT strategies. Expanding research to include other converter types—such as buck, Cuk, SEPIC, and buck-boost converters—could provide greater flexibility in system design. Conducting extensive tests with these converters would help assess and enhance the performance of MPPT approaches across different system configurations.

- **Extension of MPPT Algorithms to Wind Energy Systems**: The MPPT algorithms developed were tested exclusively on solar PV systems. Adapting these strategies for Maximum Power Point Tracking in wind energy systems offers an intriguing research opportunity. This extension could contribute to optimizing energy extraction from wind resources, promoting a more diversified renewable energy portfolio.
- **Development of Hybrid Renewable Energy Systems**: Relying solely on solar PV sources may not always ensure consistent power supply due to variability in sunlight. Combining small wind turbines with solar PV sources in hybrid systems could provide a more reliable and stable power supply to the load. Investigating the design, control, and optimization of such hybrid systems could lead to solutions that mitigate the intermittency of renewable sources.
- **Implementation of Smart Grids for Enhanced Energy Management**: Building upon hybrid systems, implementing smart grids that interconnect solar PV modules and wind turbines can significantly improve power ratings and facilitate better monitoring and management of generated and consumed energy. Research into smart grid technologies tailored for renewable energy integration is crucial for advancing sustainable and efficient energy infrastructures.
- **Advanced Energy Storage Solutions**: Exploring innovative battery technologies and energy storage systems can enhance the reliability and efficiency of standalone PV systems. Research into high-capacity, long-life, and fast-charging storage options could address the challenges of energy availability during periods of low generation.
- **Integration with Internet of Things (IoT) and Data Analytics**: Incorporating IoT devices and data analytics into PV systems can enable real-time monitoring and predictive maintenance. This integration can lead to smarter energy management strategies, optimizing performance and extending the lifespan of system components.
- **Economic and Environmental Impact Analysis**: Conducting comprehensive studies on the economic viability and environmental benefits of advanced MPPT techniques and hybrid systems can support decision-making processes. Such analyses can promote wider adoption by demonstrating cost savings and contributions to sustainability goals.

In summary, the potential for future research in optimizing standalone photovoltaic systems is vast and multifaceted. By pursuing these suggested directions, future studies can address existing challenges, enhance system efficiencies, and contribute significantly to the global transition toward sustainable energy solutions. Embracing interdisciplinary approaches and emerging technologies will be key in driving innovation and achieving breakthroughs in renewable energy optimization.

Closing Remarks

The advancements presented in this book contribute significantly to the optimization of standalone solar photovoltaic systems, promoting the viability of renewable energy solutions for remote and isolated areas. By addressing efficiency challenges through innovative MPPT strategies and exploring future research directions, we aim to support global efforts toward sustainable energy development. Continued interdisciplinary research and integration of emerging technologies will be essential in driving innovation and achieving breakthroughs in renewable energy optimization.

Appendices

A. MATLAB script of Newton-Raphson method based-modeling (Fig. 3.46)

```
%--------Characterization of Photovoltaic Panel Using Single Diode Model-
-%
%------------------- Based on Newton-Raphson Method ----------------%
       %****** Author: Ph.D.Chellakhi Abdelkhalek *****%
it=input('Enter the number of curves you aim to plot: 1,2...N: ');
for N=1:it
%----------------------------------------------------------------
-----------%
PV Module DATA
%----------------------------------------------------------------
-----------%
T=input('Enter the value of T en °C: ');         % Temperature of the celle
en °C
G=input('Enter the value of solar irradiance en °W/m2: '); % Solar Irradi-
ance en W/m2
ai=0.102/100;                                    % Current Temperature
coefficient(ki)
av=-0.36099/100;                                 % Voltage Temperature
coefficient(kv)
Isc_r=7.84;                                      % Short-circuit current
Voc_r=36.3;                                      % Open circuit voltage
Vm=29;                                           % Maximum voltage @ STC
Im=7.35;                                         % Maximum current @ STC
Pm=213.15;                                       % Maximum power @ STC
Ns=60;                                           % number of Cells
n=0.98117;                                       % Diode ideality factor
```

© The Editor(s) (if applicable) and The Author(s), under exclusive license to Springer Nature Switzerland AG 2025
A. Chellakhi and S. El Beid, *Optimizing Solar Photovoltaic Systems*, Synthesis Lectures on Renewable Energy Technologies, https://doi.org/10.1007/978-3-031-93283-0

```
%------------------------------------------------------------
-----------%
Internal parameters%
%------------------------------------------------------------
-----------%
Gr=1000;                                    % reference irradiance
T=T+273.6;                                  %
Tr= 25 +273.6;                              % Temperature reference
dT=T-Tr;
Isc=Isc_r+ai*dT;                            % variation of Isc with T°
Voc=Voc_r+av*dT;                            % variation de Vco with T°
q=1.60217646*power(10,-19);                 % charge constant
K=1.3806503*power(10,-23);                  % Boltzmann constant
Vt=(Ns*n*K*T/q);                            % Thermal voltage
Eg=1.12;                                    % gap energy
Iph=Isc*(G/Gr);                             % photo-current
Iss=(Isc)/( exp( Voc/Vt )-1 );  % saturation current
Is=Iss*( (T/Tr)^3 ) * exp ( ( (q*Eg)/(n*K) )*((1/Tr)-(1/T)) );% saturation
current
Rs=0.39383;%input('Enter the value of Rs : ');% series resistance
Rp=313.3991; %input('Enter the value of Rp : '); parallel resistance
%------------------------------------------------------------
------------%
I=Iph;                                      % initial condition
V=0:(Voc/100):Voc;                          % input voltage array
for n1=1:length(V)
   for n2=1:20                              % Newton-Raphson loop for
calculating output current
   Vd= (V(n1)+Rs*I);                        % Diode Voltage
   Id=Is*( exp(Vd/Vt) -1);                  % Diode current
   Ip=Vd/Rp;                                % Parallele resistance
Current
   f=Iph-I-Id-Ip;                           % f(I)=0
   df=-1-(Is*Rs/Vt)*exp(Vd/Vt)-(Rs/Rp);     % f'(I)=0
   I=I-f/df;                                % Newton-Raphson formula
   end                                      % end of Newton-Raphson
   if I<0
      I=0;
   end
   Ipv(n1)=I;                               % Accumulation of output
currant
end
%------------------------------------------------------------
------------%
P=Ipv.*V;                                   % Output power
```

```
figure(1)
hold on
yyaxis left                               % Creates an axes
that has a y-axis on both the left and right sides
plot(V,Ipv,'linewidth',2)
yyaxis right                              % Creates an axes
that has a y-axis on both the left and right sides
plot(V,P,'linewidth',2)
grid on
%----------------------------------------------------------------
-----------%
end
```

B. P&O MPPT algorithm M-file code in MATLAB/Simulink Platform Implementing (Fig. 3.75)

```
%************ P&O MPPT MPPT of PV Systems ****************%
    %***** Author: Ph.D.Chellakhi Abdelkhalek *****%
function Duty = P_and_O_MPPT_Algorithm(Vpv,Ipv)
Param input:
 %Initial value for D output
 Dinit = 0.35;           %Initial value for D output
deltaD = 0.005; %Increment value used to increase/decrease the duty cycle
persistent Dold Vold Pold;
dataType = 'double';
if isempty(Dold)                %Testing if the variable Dold has been
defined
Dold=Dinit;
Vold=0;
Pold=0;
end
Ppv=Vpv*Ipv;
dVpv=Vpv-Vold;
dPpv=Ppv-Pold;
   if dPpv > 0      %Test if new power is higher than old
      if dVpv > 0
           Duty = Dold + deltaD; %% Increase or decrease Duty cycle
   else
           Duty = Dold - deltaD; %% decrease Duty cycle
   end
else
   if dVpv > 0
```

```
            Duty = Dold - deltaD; %% decrease Duty cycle
    else
            Duty = Dold + deltaD; %% Increase or decrease Duty cycle
    end
end
Dold=Duty;
Vold=Vpv;
Pold=Ppv;
end
```

C. **P&O MPPT algorithm M-file code for Arduino Uno Board Implementation** (Fig. 3.82)

```
// %********************* P&O MPPT code implementation *************************%
// %***************** Author: Dr.Chellakhi Abdelkhalek *****************%
#include < LiquidCrystal.h >
LiquidCrystal lcd(12, 11, 5, 4, 3, 2);
float sensorValue1 = 0;
float sensorValue2 = 0;
float voltageValue = 0;
float currentValue = 0;
float Power_now = 0, Power_anc = 0, voltage_anc = 0;
float D = 0, delta = 0.5; //delta = 0.2
float pwm = 150;
void setup().
{
pinMode(6, OUTPUT);
lcd.begin(16, 2);
}
void loop().
{
sensorValue1 = analogRead(A0);
sensorValue2 = analogRead(A1);
voltageValue = (sensorValue1 * 5.0 / 1023.0) * 5;
currentValue = (sensorValue2 * 5.0 / 1023.0);
lcd.setCursor(0, 0);
Power_now = voltageValue * currentValue;
// ********************* LDC affiche *********************
lcd.print("P = ");
lcd.print(Power_now);
```

```
lcd.print("W D = ");
lcd.print(D);
lcd.print("%");
lcd.setCursor(0, 1);
lcd.print("V = ");
lcd.print(voltageValue);
lcd.print("V I = ");
lcd.print(currentValue);
lcd.print("A");
// ***********************************************************
// ********************* P&o Mppt ********************
if (Power_now > Power_anc).
    { if (voltageValue > voltage_anc).
         pwm = pwm - delta;
      else
         pwm = pwm + delta;
    }
else
{
if (voltageValue > voltage_anc)
         pwm = pwm + delta;
else
         pwm = pwm - delta;
}
// ***********************************************************
Power_anc = Power_now;
voltage_anc = voltageValue;
if (pwm < 15).
    pwm = 15;
if (pwm > 255).
    pwm = 150;
    D = (pwm/255)*100;         // convertir duty cycle en (%).
    analogWrite(6, pwm);
}
```

D. **INC MPPT Algorithm M-file code in MATLAB/Simulink Platform Implementing** (Fig. 3.85)

```
%************* INC MPPT MPPT of PV Systems ****************%
      %*****Author: Dr.Chellakhi Abdelkhalek *****%
```

```
function Duty = INC_MPPT_Algorithm(Vpv,Ipv)
Param input:
  %Initial value for D output
  Dinit = 0.35;                %Initial value for D output
deltaD = 0.005;                %Increment value used to increase/decrease the
duty cycle
persistent Dold Vold Pold;
dataType = 'double';
if isempty(Dold)               %Testing if the variable Dold has been defined
Dold=Dinit;
Vold=0;
Pold=0;
end
dVpv=Vpv-Vold;
dIpv=Ipv-Pold;
if(dVpv==0)
    if(dIpv==0)
       Duty=Dold;
    end
    if(dIpv>0)                              %%Test if new current is
higher than old
       Duty= Dold - deltaD; %% decrease Duty cycle
    end
  if(dIpv<0) %%Test if new current is lower than old
       Duty= Dold + deltaD; %% decrease Duty cycle
   end
  else
    if(dIpv/dVpv==-Ipv/Vpv)
       Duty=Dold;
    end
    if(dIpv/dVpv>-Ipv/Vpv)
       Duty= Dold - deltaD; %% decrease Duty cycle
    end
    if(dIpv/dVpv<-Ipv/Vpv)
       Duty= Dold + deltaD; %% decrease Duty cycle
    end
end
Dold=Duty;
Vold=Vpv;
Pold=Ppv;
end
```

E. **INC MPPT Algorithm M-file code for Arduino UNO Board Implementing** (Fig. 3.82)

```
// %********************* INC MPPT code implementation **************************%
// %****************** Author: Dr. Chellakhi Abdelkhalek ******************%
#include < LiquidCrystal.h >
LiquidCrystal lcd(12, 11, 5, 4, 3, 2);
float sensorValue1 = 0;
float sensorValue2 = 0;
float voltageValue = 0;
float currentValue = 0;
float Power_now = 0, Power_anc = 0, Current_anc = 0, Voltage_anc = 0, deltaI = 0, deltaV = 0;
float D = 0, delta = 0.5; //delta = 0.2
float pwm = 150;
void setup().
{
   pinMode(6, OUTPUT);
   lcd.begin(16, 2);
}
void loop().
{
sensorValue1 = analogRead(A0);
sensorValue2 = analogRead(A1);
voltageValue = (sensorValue1 * 5.0 /1023.0) *5;
currentValue = (sensorValue2 * 5.0 /1023.0);
lcd.setCursor(0, 0);
Power_now = voltageValue * currentValue;
// ********************* LDC affiche *********************
lcd.print("P = ");
lcd.print(Power_now);
lcd.print("W D = ");
lcd.print(D);
lcd.print("%");
lcd.setCursor(0, 1);
lcd.print("V = ");
lcd.print(voltageValue);
lcd.print("V I = ");
lcd.print(currentValue);
lcd.print("A");
```

```
// ***********************************************************
deltaI = currentValue-Current_anc;
deltaV = voltageValue-Voltage_anc;
// ********************** Inc Mppt ********************
if(deltaV = = 0)
    { if(deltaI = = 0)
      {// nothing to do
    }
  else
         { if(deltaI > 0)
             pwm = pwm-delta;
           else
             pwm = pwm + delta;
         }
    }
else
       { if((voltageValue*deltaI) + (currentValue*deltaV) = = 0)
          {// nothing to do
    }
     else
    { if((deltaI/deltaV) + (currentValue/voltageValue) > 0).
          pwm = pwm-delta;
     else
          pwm = pwm + delta;
      }
  }
// ***********************************************************
Voltage_anc = voltageValue;
Current_anc = currentValue;
Power_anc = Power_now;
 if(pwm > 250).
              pwm = 250;
 if (pwm < 15).
              pwm = 15;
          D = (pwm/255)*100;         // convertir duty cycle en (%).
analogWrite(6, pwm);
 }
```

The manufacturer's authorised representative in the EU is Springer Nature Customer Service Centre GmbH, Europaplatz 3, 69115 Heidelberg, Germany. If you have any concerns regarding our products, please contact ProductSafety@springernature.com

Printed and bound by CPI Group (UK) Ltd, Croydon, CR0 4YY

26/03/2026

02078995-0001